Democracy, Risk, and Community

ENVIRONMENTAL ETHICS AND SCIENCE POLICY SERIES
Kristin Schrader-Frechette, General Editor

ACCEPTABLE EVIDENCE
Science and Values in Risk Management
Edited by Deborah Mayo and Rachelle D. Hollander

EXPERTS IN UNCERTAINTY
Opinion and Subjective Probability in Science
Roger M. Cooke

REGULATING TOXIC SUBSTANCES
A Philosophy of Science and the Law
Carl F. Cranor

IN NATURE'S INTERESTS?
Interests, Animal Rights, and Environmental Ethics
Gary E. Varner

DEMOCRACY, RISK, AND COMMUNITY
Technological Hazards and the Evolution of Liberalism
Richard P. Hiskes

Democracy, Risk, and Community

Technological Hazards and the Evolution of Liberalism

Richard P. Hiskes

New York Oxford
OXFORD UNIVERSITY PRESS
1998

Oxford University Press

Oxford New York
Athens Auckland Bangkok Bogotá Buenos Aires Calcutta
Cape Town Chennai Dar es Salaam Delhi Florence Hong Kong Istanbul
Karachi Kuala Lumpur Madrid Melbourne Mexico City Mumbai
Nairobi Paris Singapore Taipei Tokyo Toronto Warsaw

and associated companies in
Berlin Ibadan

Published by Oxford University Press, Inc.
198 Madison Avenue, New York, New York 10016

Oxford is a registered trademark of Oxford University Press

Library of Congress Cataloging-in-Publication Data
Hiskes, Richard P., 1951–
Democracy, risk, and community : technological hazards and the
evolution of liberalism / Richard P. Hiskes.
p. cm. — (Environmental ethics and science policy)
Includes bibliographical references and index.
ISBN 0-19-512008-6
1. Technology—Risk assessment. 2. Technology—Social aspects.
3. Technology and state. I. Title. II. Series.
T174.5.H57 1998
363.1—dc21 97-35724

1 3 5 7 9 8 6 4 2
Ptinted in the United States of America
on acid-free paper

For Anne and Ben,
two sure things.

Acknowledgments

Anyone who writes for publication already possesses an intuitive grasp of two of the predominant themes of this book. First, this book explores the idea and impact of risk, and authors recognize that all professional writing exposes them to risks: intellectual, psychic, and, these days if one writes on computer, the special technological terror that comes from forgetting to "back up one's hard drive"—the mantra of 1990s authors. Second, like ideas themselves, all books are intrinsically "emergent" phenomena, whether the author is willing to own up to his intellectual debts or not.

These are the pages where authors usually admit their book's emergent status, so I will cheerfully remind all potential critics to review this and their own books' acknowledgment pages before they dismiss my underlying contention that phenomena exist that we all recognize as solely the products of collective rather than individual events or actors. Risk, I will argue throughout this book, is such a phenomenon, whose emergent nature even the most solitary of writers must sooner or later acknowledge.

Having thus anticipated (and tried to defuse) criticism, I want especially to thank those who lessened the riskiness of this particular phenomenon. First and foremost, Professor Kristin Shrader-Frechette, the editor of the series of which this book is a part, was present at the very start of the project, and her special abilities to prod, cajole, and encourage kept me believing (and working) whenever the risks seemed ready to overwhelm. She is also a foremost scholar in the field of risk analysis, and her substantive comments on the manuscript are present behind every page. Professor Larry May also deserves my heartiest gratitude for reading the manuscript twice in its entirety and offering his own critique. The present version would not have been possible without his ideas, willingly shared and duly incorporated.

Many friends, students, and colleagues at the University of Connecticut and elsewhere also had a hand in the emergence of this book. Professor Amy Gutmann of Princeton offered commentary and valuable insights on chapter six. George Graham at Vanderbilt offered a supporting voice in this endeavor, as in so many others throughout my career. Colleagues Larry Bowman, J. Garry Clifford, John Rourke, Ernie Zirakzadeh, and Christopher Rocco offered support ranging from a friendship willing even to overlook egregious errors on golf scorecards to intellectual critiques willing to point out equally glaring errors in the manuscript. Many lively conversations with my colleague Elizabeth Hanson anticipated several themes incorporated here, and frequent discussions with my learned friend and environmental risk specialist Randy May provided a needed sobering effect on my enthusiasm for nonspecialized input into risk decisions.

The University of Connecticut underwrote the research and writing of this book in a variety of ways. Dean Timothy J. Killeen provided equipment support, Associate Provost Thomas Giolas made space available both in his schedule and his office domain for the work to proceed, and the Provost's Research Fellowship gave me the necessary time away from teaching to complete the initial stages of the project.

Students also had a major hand in the development of the arguments presented here, even beyond that which takes place in classes and graduate seminars. Kathy Flavin and Professor Jessica J. Kulynych read large portions of the manuscript and commented heavily upon them. Hwang Sung-Uk also provided insightful criticism, as did James Harrigan, who is heartily sick of encountering various versions of "Sunday Afternoon in the State of Nature." Thanks to all for their perseverance and patience.

Several portions of this book were presented at annual meetings of the American Political Science Association, and I would like to thank participants on panels and in the audience for their suggestions. Anne Deckard offered research assistance and called my attention to relevant scientific literature. Finally, to my wife, Anne, and son, Ben, I extend my deepest thanks and dedicate the book: Anne for her willingness to share her brilliance as a philosopher as well as her life; and Ben, whose own emergence shortly before page one gave new meaning to old words—risk, responsibility, joy.

Contents

Democracy, Risk, and Community

Introduction

Two concepts dominate the political consciousness and even the popular culture of many societies today. They are ideas sharing an association at least since the seventeenth century and arguably backward to the time of the ancient Greeks. Today their individual definitions rely, I argue, on a curious symbiosis, and both their definitions (and therefore their relationship) are currently in a state of evolution and flux. They are the concepts of democracy and risk. The following chapters explore their present interrelationship as a defining aspect of modern political and social culture.

Risk and democracy since the seventeenth century have shared a common heritage within the political theory of liberalism, which views both as products of individual behavior. Democracy in the liberal mode is described largely in terms of protected features of individual life such as liberty, privacy, and rights of participation. Similarly, liberalism also views risks as the products of individual choices and behaviors; historic examples include the decisions to explore the "New World," or to "go West." More contemporary risks, viewed as liberalism sees them, involve taking individual economic gambles or risking one's present or future welfare in the pursuit of excellence, pleasure, or gain of one sort or another. Today, however, it is increasingly difficult to view such pursuits as strictly solitary acts about which no one else has an interest or stake except the individual. It is the organizing thesis of this book that the ontological status of risk has been fundamentally altered by the rise of modern technological society and, as a result so, too has life in democratic societies.

Unlike risks perceived through the liberal lens of individual decision and behavior, the risks that arise from modern technologies are in some important ways not reducible to either the individual decisions, actions, or people participating in them. Consider the risks arising from technologies like modern communications, data processing, or nuclear energy; from behaviors like smoking or unprotected sex; or from political decisions such as those involved in siting hazardous installations. They share the same basic definition of all other risks as "exposure to the chance of loss."[1] Yet I contend that in a fundamental way these risks are different in that they are "emergent" phenomena—the results of collective decisions or behaviors not completely reducible in a meaningful way to the individual people involved in them. As I define them at length in the first chapter, emergent risks have collective histories—partly contained within the history of science taken as a community, partly within political history, and partly within their own policy histories. They also have futures, futures that are disconcertingly independent to some extent of how individuals individually choose to cope with them: nuclear waste will persist in some of its forms for millennia; the personal risks of tobacco use persist through the dangers of passive smoke whether one quits smoking or not; moving to the country can land one in proximity to society's waste dumps and breathing somebody else's polluted air as it drifts in from cities in distant states.

Emergent risks, I argue, call attention to our lives *together*, within a community of interconnected beings whose actions pose consequences—even risks—for each other. Traditionally, liberal politics has called attention to our lives *apart* as separate political beings with discrete interests protected within democratically defined cocoons of private rights and liberties. The emergent, intersubjective character of risks threatens forcibly to open those cocoons, exposing the individuals within to the risks that exist as irreducible and inescapable consequences of modern social life. Because of their emergent character, risks today are intrinsically befuddling to liberal societies, whose governing notions circulate around a seventeenth-century ontology of reductionistic political ideas such as individual rights and around a set of mechanistically arranged political institutions separately checking and balancing each other's individual actions. Thus, liberal societies experience great difficulty in making policy for modern risks, and this inability places them and their citizens in additional jeopardy. Therefore, the following argument contends that unless a transformation takes place outside the cocoons of liberal privacy, that is, within the public realm of policy making for risk, the risks we collectively face will continue to fester and infect our private lives as well.

I have much more to say by way of definition and exploration about the emergent nature of risk in the chapters that follow. But putting aside temporarily the disputes within the philosophy of social science in the 1950s concerning the status of emergent phenomena, let me begin with an example of them unique to our technological times.

When I was growing up in Chicago in those 1950s, I recall listening to rush hour traffic reports over my father's car radio, in which the reporter aloft in a helicopter over one of the city's many busy expressways would often warn listeners of a traffic delay due to a "gapers' block." What he meant, of course, was a traffic stoppage resulting from drivers slowing down to gawk at an accident. Implicit in his name for this traffic phenomenon was both blame and an attribution of cause for the slowness. I remember times when, caught in a gapers' block, and, in fact, doing our share of gaping ourselves, my father would bellow at the radio voice that *he* was not the problem, it was all the other drivers in their sick need to view highway carnage that were causing the delay.

Phenomena like gapers' blocks are not very mysterious, since of course they are the result of the actions of individuals, specifically those involved in the individual event of the accident itself, and of all the drivers who slow down as they go by. Yet there is also something ineluctably, even comically, collective about such events. The problem is not that one person is slowing down to gape, but that so many are, and that the collective effect seems so difficult to avoid. By definition, the event of the block requires a multiplication of individual actions, but within the context of traffic the effect appears so singular, the result not of many disconnected drivers but of a "chain" of events and individuals.

Gapers' blocks are in one sense simply the products of a set of individual behaviors, but they also witness to the special, sometimes unnerving connectedness of our technological age. It is hard to imagine a gapers' block created by pedestrians or horseback riders in an earlier century. Also, individuals within the privacy of automobiles becalmed in gapers' blocks tend, like my father, not to see themselves as in fact the problem—someone else's morbid curiosity is to blame. Finally, the solution to events like gapers' blocks is not really to try to change individual behavior. Instead, the solution lies in a collective decision and action—widening the road or providing (through public funds) for more efficient emergency staff to move traffic more quickly. Gapers' blocks are not inevitable any more than risks are "autonomous," as they are sometimes erroneously viewed. But they are events that are uniquely part of our technologically developed and risky times, and coping with them requires different conceptualizations than those provided by liberal politics and social science.

This is a book about conceptual change and about how its impact reverberates through the practical politics of liberal democratic societies, demanding changes in that politics and in the conceptualizations that underlie it. As the nature (and, as I begin in chapter 1, the perception) of risk is different today than in periods before the advent of modern technologies, its impact upon concepts that underlie liberal society is just beginning to be felt. But where it is felt, the policy reverberations are unmistakable: controversy, fear, anger, the "not in my backyard" (NIMBY) syndrome, stalemate, and, finally, inaction. In order to understand the very much real-world policy problems surrounding technological risk, it is necessary to explore the interaction of risk at the conceptual level with basic concepts that define the liberal attitudes and expectations of modern citizens. In this book I seek to prepare the way for effective and democratic policy making for risk by identifying where risk challenges liberal understandings of some of our most precious liberal ideas: liberty, consent, legitimate authority, community, rights, and responsibility. As risk affects the meanings of all of those concepts, it upsets the policy process for risk by rendering uncertain the criteria for political decision, criteria identified for liberalism centuries ago but still embraced by citizens today.

Political theory is comfortable operating at the level of concepts, of course, and theorists assume easily (perhaps too easily) that concepts interact with and produce real-world consequences. Certainly we have witnessed within the last decade other examples where concepts like liberty and democracy have wrought remarkable political upheavals in parts of the world not previously considered hospitable to either idea.[2] Yet to speak of the interplay of risk and other political ideas is faintly disturbing for two reasons. The first is applicable whenever conceptual change takes place, for, as Alasdair MacIntyre observes, "to alter concepts, whether by modifying existing concepts or by making new concepts available or by destroying old ones, is to alter behavior."[3] If it is risk that is forcing the change in our concepts and therefore our behavior, our present sensation of not being in complete control of modern risk adds a special feeling of helplessness in the face of a new political reality. As risk causes conceptual changes, it also then, as all conceptual changes portend, represents political innovation,[4] in this case innovation that is the product of an emergent process to which liberals almost by nature are blind. Thus, the risk-induced alterations of our political concepts and practices are more disconcerting than most such changes; they take us by surprise.

The second cause of unease in the interaction of risks and political concepts arises from the scientific paternity that technological risks share. Science presumes a determined world within which explanation and prediction are recognizable goals, but we are uncomfortable about allowing such determinism into our political consciousness. Risk as a product of scientific and technological development may be inevitable in some sense, but to say that it also inevitably will change the way we think and act about political things is to violate at a visceral level the liberal belief that liberty, even free will, is possible. It introduces a disconcerting level of determinism into everyday life. Emergent risks, like gapers' blocks, make us feel that we do not control our environment or the events of our own lives as much as we might like to think. As one of liberalism's most eloquent and influential voices, Sir Isaiah Berlin sounds an alarm worth heeding again before we proceed further:

> If social determinism is true, and we begin to take it seriously, then the changes in the whole of our language, our moral terminology, our attitudes toward one another, our views of history, of society, and of everything else will be too profound to be adumbrated. The concepts of praise and blame, innocence and guilt, and responsibility from which we started are but a small element in the structures which would collapse or disappear.[5]

I am aware of the charge of determinism that exploring the relationship between technological risk and political concepts might well level against me. But, following the pragmatists in general and John Dewey in particular, I believe it important to realize that all concepts are grounded both in experience and, more importantly still, in action, and that to expect any different is to await a certainty not forthcoming from either science or politics.[6] As we will see throughout this book, the technological products of science have cast doubt upon many traditional liberal conceptions of epistemology, morality, and politics. Yet this need not make us accept the dispiriting and fallacious conclusion that somehow technology is itself autonomous and determinative of our political, moral, even spiritual life. What modern risks do reveal, however, is that, though not autonomously produced, technological effects are not reducible to the individual persons, events, or decisions incorporated in them. Risks are *not* autonomous and do not of themselves determine our lives; they only appear so if we fail to see that they are the emergent products of our lives together—lives in connection with others, where those connections are both new and different products of modern technology.

We should not belittle Berlin's fears of technological determinism, however, especially since chapter five is largely devoted to exploring exactly how concepts of "praise and blame, innocence and guilt, and responsibility" are, in fact, altered severely by the impact of technological risks. In a sense Berlin is right—as the following chapters make clear, emergent technological risk is forcing changes in the intellectual and normative structure of life in modern liberal societies that are indeed profound, so much so that making policy for them is stymied at almost every turn, even though such indecision runs the further risk in some instances, such as radioactive waste disposal, of unthinkable and unknowable catastrophe.

The recognition that by its impact on relevant political concepts risk causes such policy paralysis not only helps to explain such new phenomena as the NIMBY syndrome, it also points the way toward effective response to the practical changes that risks herald within the political practices of liberal societies. That is, understanding the conceptual and therefore the political changes wrought by technological risk is a guide for action in how to cope with them. We explore this perhaps unexpectedly upbeat consequence in chapter six, when discussing how democratic practices must innovate in response to the challenge that emergent risks pose.

Democracy and risk have always been affiliated concepts. Thus, it should come as no surprise that part of the response to the threats that technological risks present has been an upsurge in political participation by citizens who stand to be most affected by new risks. Perhaps the NIMBY syndrome deserves to be considered, on one level at least, as the late-twentieth-century manifestation of the Jeffersonian requirement of a little democratic revolution every few years. Risks have reinvigorated democratic participation to some extent, in other words, regardless of what we think of the motivations behind such organized responses to policy initiatives. This participatory legacy of contemporary risks should not surprise us once we learn to view them ontologically as irreducibly emergent, collective phenomena. It is, after all, an ontology risks share only with democratic political systems. Democracy too can only be realized in its emergent form. It is the product only and always of the joint activity of citizens—not of only one or several of them, as is the case in monarchies or aristocratic forms of governance. Thus, uncovering the democratic requirements of successful policy making for contemporary risks will itself be an act of democratic political innovation.

Since this is a book about concepts and how they change, it is important before going further to be as clear as we can about what those concepts signify prior to their interactive transformation. Toward that end, each of the following chapters begins with an exploration of how concepts such as risk, consent, authority, rights, responsibility, and community have traditionally been construed within the literature of liberalism, in the absence of awareness of risk's impact. Following that, each chapter does three things. First, it elaborates how technological risk has affected the accepted meaning of concepts, whether or not either citizens or commentators have yet recognized those impacts. Second, each chapter explores how risk opens the way toward redefinition of liberal concepts in light of their interaction with risk. In some cases this is more speculative than others, since what the "community" of the future will look like is not easily discernible, and in some cases the redefinition may no longer be recognizable as liberal. Finally, each chapter anticipates how the impact of risk on that chapter's conceptual focus warrants changes in policy making for risk, always with an eye to the final chapter's discussion of how to maintain the imperatives of democracy within the decision process.

Throughout the following chapters, reference is often made to "liberalism," "liberal societies," and, generally, to things "liberal." Since liberalism these days is itself a disputed concept, and logic would seem to dictate that we will need to recognize what is being threatened if we have any hope of recognizing the threat, I need to say a few words about how I am using liberalism as a backdrop to the discussion of risk and political concepts.

Without sounding too cagey, much of what I take "liberalism" to indicate will "emerge" in each chapter's discussion of different liberal hallmarks. That is to say, understanding liberalism requires separate explorations into the traditional and now threatened constructions of its central concepts. Nevertheless, I should at least list here those central concepts of liberalism and their place in what follows: consent of the governed (chapter 2), legitimate—and therefore limited—political authority (chapter 3), individual rights (chapter 4), moral responsibility (chapter 5), and the role of democratic participation in the identity of liberal citizens (chapter 6).

Many scholarly treatises have been written with the express purpose of defining, exposing, and otherwise criticizing or eulogizing liberalism, and I will refer to a variety of these works in the following chapters. Yet I want to call attention as a final preparation to what follows to two universally accepted aspects of liberal political theory. Liberalism, first, always focuses on the individual as the unit of political analysis, and, second, always draws its (abstract) picture of that political entity as living within two separate spheres of activity, interest, and involvement—the public and private realms. I highlight these aspects of the liberal creed especially, because it is these two that are most challenged by the emergent nature of contemporary technological risks.

The understanding of risks, as they confront us today, is not amenable to the reductionist political analysis of individuals and their isolated behaviors presumed by liberal politics and liberal social science. Risks are the product of our lives together; furthermore, they cannot be sequestered into either the public or private sides of the lives of liberal citizens. Though they often are surrounded by public controversy and fill the political agenda of modern democracies, modern risks cannot be escaped once we leave the political realm and return to the privacy of our own homes. Modern technologies invade our homes and bring their risks with them, forcing their way into our private spaces through our devices of communication, our technologies of entertainment, reproduction, comfort, nutrition, and elimination of waste. The risks these technologies deliver are in the air, the food, the food chain. How risks got there was never only a matter of individual or private choice; getting rid of them cannot be a matter solely of individual effort. Liberalism, like liberal citizens, is bewildered by modern risk, but liberalism has proven itself an adaptable political doctrine. It must be so again. As the following discussion demonstrates, modern risks make imperative the evolution of liberalism as a matter of its—and its citizens'—survival, in every sense.

1

Risk and the Redefinition of Politics

The three concepts in the title of this book could be juxtaposed in a variety of ways yielding widely different meanings. "The Democratic Community at Risk" darkly presages a threat to communities or to their democratic constitutions. "The Risk of Democratic Community" seems a warning about either the practices of democracy or the ties of community. "Community Assumes Risks of Democracy" trumpets a political headline of our time, in which all three elements seem laudatory, heroic, comforting. Democracy, risk, and community are essentially contested ideas of today's political world, and the possible permutations in their juxtaposition go far in setting the parameters of contemporary political discourse. The meaning of these terms both individually and in combination is often at the heart of current political disputes between political leaders, between leaders and citizens, and between citizens themselves. From issues of war and peace to disputes about land use and "the conquest of bread," politics today demarcates the boundaries of communities, distributes risk, and claims procedures to be democratic.

The purpose of this book is to explore the relationship between these perplexingly rich ideas within contemporary politics and political theory. I am specifically interested in what is often generically called "technological" risk and how it affects political definitions of democracy and community. It is my contention that, when compared to risks characterizing earlier historical epochs, today technological risk has a special—or at least new—character. Furthermore, this changed meaning of risk plays particular havoc with accepted definitions of democracy and with democracy's constituent political concepts and practices, especially when democracy is construed liberally.[1] In other words, the existence and perception of risk threaten to alter our political decision-making processes fundamentally, as well as our understanding of the concepts of authority, rights, consent, and liberty, among others, that those processes are presumed, in a democracy, to manifest.

The idea of risk has historically been linked with the urge toward democracy, of course, beginning with the ancient Greeks. Plato and Aristotle certainly agreed that a democratic reliance on the ability of common people to govern both themselves and a nation as a whole was too risky a proposition to pursue. Beginning in the seventeenth century, however, the risks of democracy became associated with classical liberal ideas of individual liberty and republican politics as enunciated by Hobbes, Locke, and the American founders.[2] Democracy might still be risky, but the goals of individual liberty and equality made it a risk worth taking. It is this association of risk with liberal democratic politics that is my focus here. My contention is that the nature of technological risk threatens to upset this historic marriage of risk taking and democratic liberty within liberal democracies like the United States.

Liberalism, at least till the mid-twentieth century, embraced the ideas of risk and risk taking as part of its promise of individual autonomy and responsibility. Indeed, as I argue in chapter four, one of the great documents of liberalism—the US Constitution—guarantees in its first 10 amendments the liberty to take risks as a matter of right. Partly as a consequence, especially in the United States, the convergence of liberal politics with a free market economy valorized risk taking as essential (and essentially) within the ethos of individual freedom. Capitalism, of course, attempts the merger of liberty and risk taking by extending the promise of future reward to those willing to run the risks of laying it all on the line. When wedded to the siren calls of the "New World" or "Going West," this pact with the future ensured that citizens of the United States have an intimate relationship with risk that is bred into their cultural legacy.

Today, some critics argue that America's unique and uniquely productive relationship with risk is at an end.[3] Cries of "not in my back yard" do seem to ring out across the land, and in many cases have succeeded in blocking policy concerning the siting of technological installations of various types.[4] This growing rejection of risk is something we will be much concerned with in this book, but to understand it we must not assume with some that the United States (as well as other technologically advanced societies) has become a nation of wimps. What has changed is the nature of risk itself; it has mutated into a form fundamentally at odds with the liberal heritage of most US citizens. It is risk without the adventure of the frontier or the financial return of capitalist entrepeneurship. It is risk not defined privately but interpersonally, a kind of threat that individuals do not consciously take on or accept but gradually find themselves enduring; it is risk not identified with individual persons or actions but emergent at the level of social life and collective choice. This is a form of risk not easily associated with liberalism or democracy. It threatens both, because it brings liberal democracy face to face with a value and social fact it has ignored or relegated to the second tier of political virtues for three centuries: the idea of community.

To say that risk must be viewed in a communitarian sense is to highlight two aspects of its nature that have not been emphasized sufficiently either within the relevant literature about technological risk or in society at large. They are, first, that risk is an *emergent* social fact and, second, that it is an intrinsically *political* concept and phenomenon. The remainder of this chapter explores what these two attributes of technological risk mean and how they affect scholarly and policy disputes about what risk is and how to make policy decisions for it. To see risks as emergent and political is to see them in a new light, a light that casts a shadow on classical liberalism and its attempts to relegate risk activity to the realms of private choice and individual political action. The remaining chapters seek then to illuminate the shadowy concepts of liberalism that will need revision if democracy is to cope with its new risk environment.

The inclusion of risk in conjunction with democracy and community in the title is meant to call attention to their similar ontologies. We have little trouble seeing that democracies and communities are collective concepts denoting compound entities. Furthermore, they clearly embrace features beyond merely the collected attributes of the individuals within them; therefore, to understand either a democracy or community we need to do more than look at the isolated actions of their members. The same is true, I contend, for modern risks. Like the "gapers' blocks" of the introduction, risks present features not reducible in a meaningful way to the individual persons or events that bring them about. Moreover, even if we deny the explanatory or causal emergentism of risks, coping with them as the objects of policy making and political struggle makes clear their emergent character.

The argument in this chapter seeks to elaborate the new, political nature of technological risk under four headings featuring four different aspects of risk's nature and politics. First, in the largest section that follows, I explore what it means to view risks as emergent. Seeing risks in this new way requires, first, that we understand that the dispute between realist and culture theorists of risk concerning the connection between actual and perceived risk is most productively approached with the recognition that as emergent phenomena, risks are best viewed at the level of groups. At this level, both realists and culture theorists have important contributions to make to our understanding of emergent risks. This section also explores the role of rationality, culture, and politics (seen as the domain of interest and power) in our definition of risks as emergent phenomena.

Following this sustained ontological and epistemological exploration into risk's emergent and political nature, three other aspects of the politics of risk are discussed in later sections. First, the pervasiveness of risk in modern society ensures that the politics of risk also extend far beyond the realm of policy makers and political institutions. As a political phenomenon, risk manifests conflicts and exercises of power visible in all aspects of modern social life. Second, the pervasiveness and emergent qualities of risk hint at a possible connection between risk and democracy that warrants further exploration. Here we will see that as an emergent and political thing, risk both presumes and threatens democratic politics, an ambivalence that only a newly conceptualized democratic political realm can mediate. Finally, the conclusion of this chapter comes to grips with the disturbingly illiberal character of modern risk, when correctly apprehended in its emergent, pervasive, and political garb. Because it is fundamentally unrecognizable by liberal social science or liberal politics, modern technological risk requires new conceptualizations of central political and moral concepts if we are to cope successfully with its impacts in our everyday life. Those conceptualizations are the focus of each of the remaining chapters.

The Emergent Politics of Technological Risk

Before exploring the newly emergent and political sides of risk, let us clarify what technological risk means. As we will shortly see, the concept of risk is beset by a number of terminological disputes, but even if we begin at the most basic level by accepting that risk simpliciter refers to "the chancing of negativity—of some loss or harm,"[5] adding the qualifier "technological" immediately opens several alternatives for definition. First, we might simply recite the risks of modern technological society, with its attendant pollution problems, threats of irradiation, occupational hazards, transportation dangers, and the like. Some of these risks are immediately recognizable as unique to the twentieth century, or at least to "industrial" society, since they are the direct result of technological innovations and development. Second, we should include risks perhaps present before our technological age but were not recognized due to lack of knowledge. Health risks are obvious examples here. Also, some such risks perhaps unknown but present in earlier times might require a further precipitating cause characteristic of modern society. Secondhand smoke is an interesting example here, for the risk it presents is compounded by modern urbanization and mass transit technologies, among others. It is difficult sometimes—culture theorists would say impossible—with these second kinds of risks to separate conceptually the risk itself from its perception, a definitional conundrum of risk to which we will shortly return.

Third, we also recognize risks that come directly from the acquisition of some new knowledge itself, whether arcane or widely held. Such risks threaten psychological or emotional harm or pain and would not have been possible before the knowledge or delivery technique was attainable. An example here stems from genetic and reproductive technologies that, when used, produce facts about persons or fetuses that subsequently require painful personal or social decisions. Such knowledge can also lead to illicit or unfair practices, as in hiring situations following genetic screening of job applicants. In these situations one more risk is produced: the risk to one's own (or society's) sense of justice or morality.

These three approaches are not exhaustive by any means but jointly demonstrate that modern technological risks threaten us in three ways: physically, psychologically, and morally.[6] These risks are all related to each other also, somewhat obviously, in that they are the products of the implementation of increased scientific (or simply technical) knowledge about our world and ourselves. None of this is very controversial and indeed has been said before by a host of commentators. But if these risks are the result of the accretion of scientific knowledge, then the method by which that knowledge is acquired gives us a clue to a unique feature of the risks we live with today.

Philosophers of science tell us that the growth of scientific knowledge is the result of a communal process. That is, no one scientist "discovers" knowledge alone but builds his or her hypotheses on the work of others and awaits confirmation of the results through the further research of fellow scientists. Philosophers may disagree whether the process of confirmation consists of replication, falsification, or political struggle,[7] but the fact remains that the search is a collective one, and knowledge emerges from interactions within the scientific community.

What I want to suggest is that an analogous process gives rise to modern technological risk. Modern risks are the result of the increase and use of scientific knowledge, and just as that knowledge "emerges" from the interactions of players within an epistemological process, so risk emerges from the interstices between individuals within the common life of modern society. This parallel should not be taken literally as a move to blame science for modern risk. Rather, the sociology of science presents us with an illustration or metaphor for how to view technological risk as a whole. Just as the logic of discovery cannot be reduced to the mind or work of a solitary scientist, the risks that characterize life in modern technological societies cannot be reduced solely to the actions or decisions of individual persons. Risk is the product of our lives *together*, and to fully understand risk's emergent character is to realize that most of the efforts to either explain risk or cope with it within an individualistic political framework are doomed to failure because they do not acknowledge the "togetherness" of our risky present.

From the standpoint of social science, my approach means at the very least that its dominant contemporary method of discussing risk has been fundamentally flawed. Rational choice theory, even in its most social and least atomistic formulations, places risk in too narrow a context to account for its emergent character.

I am aware of the battles fought within the philosophy of social science in the 1950s and after concerning the status of emergent qualities. I do not want to resuscitate that entire dispute beginning with Emile Durkheim and engaging May Brodbeck, Ernest Gellner, and others,[8] nor do I want to be subjected to the old implication that accepting the emergent nature of social facts invariably means a Marxist-leaning ontology is in use. It is true that Marx's definitions of labor as "socially necessary" and of true consciousness as "class consciousness," among his other concepts, are indeed emergent definitions, but by itself this fact obviously does not mean all emergentist conceptualizations are Marxist in origin. Similarly, there are degrees of emergentism, and my approach to risk manifests a less than total reliance on the emergentist features of risk as a phenomenon.

I do not deny, for instance, what individualists and rational choice theorists accuse emergentists of denying: that risks are open to scientific explanation or prediction. Rather, my concern is that if a correct approach requires that we view risks in an excessively methodologically individualist manner, we will defeat our purpose in wanting to explain or predict risks in the first place. We need to *cope* with risk, and coping with risks requires that we understand two things about them. First, given the complexity of life in technologically advanced societies, arriving at rigorously nonemergentist causal explanations of risk is extremely difficult, even if not actually impossible. Second, politically such an attempt is foolhardy, even dangerous. Making effective policy for risk requires that we leave behind the excessively rationalistic and methodologically individualistic presumptions that stand behind liberal politics. This second position will be elaborated throughout the chapters that follow, in which I argue that the resulting communalistic approach to risk is also more amenable to the true nature of democratic politics.

All of this makes my approach what Richard Rudner calls one of "*relative* emergence" (emphasis Rudner's).[9] Such a thesis argues only that "relative to certain circumstances, it is not *technically possible* to predict the occurrence of certain events" (emphasis Rudner's). My view is even weaker, since I claim only that it is unfruitful both epistemologically and politically to seek a nonemergentist view of risk, not that it is technically impossible. Rudner considers such a relatively emergentist explanation of risk "doubtlessly true." More than this, I consider it politically essential.

Paul Slovic somewhat anticipates this new approach to risk when commenting upon the speed with which potentially risky discoveries make their way into the lives of millions of people. The artificial sweetener Aspartame is now consumed by hundreds of millions of people daily, Slovic points out; thus, if it turns out that some hazard has been overlooked in the approval of this chemical—as in the case of the drug Thalidomide, "the potential for harm is enormous."[10] Such a degree of harm is directly related to the extraordinary complexity and interconnectedness of modern technological systems, within which it is difficult if not impossible to isolate a single causal agent. It is true that we might indeed be able to identify the inventor or discoverer of Aspartame and easily name the members of the Food and Drug Administration responsible for the chemical's approval, but, as we will see in chapter five, responsibility and even agency for such decisions ultimately lies with the wider culture itself—describable in this instance as obsessed with calories and the chimera of youthful slimness.

Although Slovic views the nonreducible interconnectedness of modern risk as a rather minor characteristic of its nature, other scholars see risk in its modern technological form as the very definition of modernity. Chief among them is Ulrich Beck, whose literary phenomenon, *Risk Society*,[11] paints a picture of modernity as sitting atop the "volcano of civilization." In his view, our threatened condition derives from Beck's characterization of modern society as constructed entirely upon relations aimed at the production and distribution of risk. Importantly, those relations are not individually isolatable or describable in terms of discrete individuals or events. Risks arise from the historical nexus of capitalism, modern science, and political liberalism and must be viewed as social products, much as Marx once viewed wealth.[12]

Beck probably goes too far in his reduction of all modern social relations to the dynamics of risk production and distribution. But in his hyperbole lies an important realization—that modern risks are both consequence and continuing (in Beck's terminology "reflexive") cause of the extraordinary degree of technological interdependence that defines modern social life. That degree of interdependence for Beck makes for a harrowing future of trying to cope with risk, especially since the unintended and unpredicted nature of modernity's riskiness makes its producers—modern science and instrumental rationality—more or less powerless in providing assistance.

Beck's explicit characterization of the rationalistic approaches to risk as particularly impotent demonstrates a basic similarity between his approach and my own rejection of rational choice models of risk assessment and policy making. Although I reject Beck's excessive pessimism, an essential insight of his analysis remains crucial to any adequate understanding of risk today, even though it is unstated in his work. An implicit sine qua non of his position presumes that risks are not reducible or isolatable to individual components (actors, events, policies, etc.) within the vast interconnections of modern life. Rather, they are visible only as they emerge from that life and reflexively affect it.

The emergent character of modern technological risks indicates several features, each of which evokes a current controversy within the literature of risk. First, it means that risk is a phenomenon that occurs at the level of groups. Second, to see risk as an emergent feature of modern society is to recognize its twofold character as an objectively verifiable fact of the external world and as a subjective response to that world. The dispute still rages between risk theorists who point out the two sides of risk as objective and subjective and choose one or the other as the most significant,[13] but the emergent, group-relatedness of risk adds a new wrinkle. If risk is not reducible to individuals, then the subjective beliefs and values that play a role either in the assessment or management of risk (or both, depending on one's position in the ontological dispute about risk), and which are evident in individuals' responses to risk, are themselves a part of the ideological makeup of a society as a whole—its beliefs and values. Thus, at the very least, the study of individuals' responses to risk must view individuals as products of their culture, a culture in some (large) measure defined by the existence of risk.

Third, in any society cultural values change, ideologies evolve, and beliefs about the world are altered as more knowledge is acquired. This evolution is a feature of the growth and development of scientific knowledge as well as the risks generated from it. These evolutionary developments in values, knowledge, science, and risk all come about through the exercise of some kind of power. That power may be in the form of persuasion, moral or otherwise; it may be in the guise of physical force or legal coercion; it may be yet more subtle. The power to define risk and attitudes toward it is a power widely dispersed in modern society, exercised by governments, media, corporations, interest groups, churches, and others. As part of culture, the definition of risk is always a product of the exercise of power. It is this cultural changeability that makes risk so intrinsically a political phenomenon and, as Roger E. Kasperson notes, ensures that "the definition of risk is essentially a political act."[14] The power to identify risk and our attitudes toward it mirrors the distribution of power in society generally, a distribution that is pervasive, amorphous, elusive, irresistible. Let us explore further the interplay of power with these three aspects of emergent technological risks.

Risks, Perceived Risks, and Groups

George L. Priest observes that from "the standpoint of the control of risk, it is difficult to identify a truly solitary act."[15] Because we are all interested in controlling risk, this lack of discreteness poses several problems. But what exactly does it mean to imply that individuals are somehow invisible in the risk equation? This becomes an especially vexing problem if one amends Priest's sentiment to include (as culture theorists do) the perception, even creation, of risk in addition to its control. Surely individuals still confront risks, take risks, suffer the consequences of risks in terms of personal gain or pain. So what does it mean to say that technological risk is a phenomenon that emerges at the level of groups?

Consider the risks that arise from technologies like modern communications, genetic screening, or nuclear energy; from behaviors like smoking or engaging in unprotected sex; or from political decisions such as siting policy. The results (that is, the risks) that emerge from all three are the products of groups of individuals difficult to recognize except as groups. Furthermore, the resultant risks are identifiable mainly in group terms; that is, they are predictable only as probable effects for certain populations, although any one individual member of a targeted population may not be affected.

To take a specific example: nuclear energy was developed within the scientific community functioning as a community. Policy for nuclear energy, such as it is, was generated gradually since the Eisenhower administration from a variety of policy sources both within and outside government. The effects of nuclear power, both positive from its proper functioning and negative from occasional (and potential) breakdowns, are seen in large-scale terms—in the energy demands and usage of society and in differentiated effects of irradiation within populations in proximity to a faulty (e.g., leaking) nuclear plant. In short, if we wanted to know who was the source of nuclear technology, who to hold responsible for the development (or lack) of nuclear policy, or who manifests the positive and negative effects of the technology and its related policy, how could we isolate the relevant actions or identify all the individually relevant actors? Many have had a role in the development of the technology and therefore in the dissemination of its risks. However, in a very real sense it is our culture's demand for ever increasing levels of power that provided the motive force for development. Knowing this, we can view the emergence of nuclear technology in either of two ways. We can see the technology itself as "autonomous," in Langdon Winner's famous reformulation of Jacques Ellul,[16] or we can stop trying to reduce all phenomena to their individual parts. The paranoia and anthropomorphism of the first proposal recommend the second. But in accepting the nonreducible nature of risk we must realize that in the search for safety not all risks have easily identifiable causes or agents. Furthermore, the whole idea of agency, as we will see in chapter five, itself becomes a bit murky. The emergence of some risks defies our use even of active verbs—like the dangers of "passive smoking," the action implied by some risks can only be described in passive voice.[17]

In a sense, the emergent nature of risk simply calls attention to our lives together within a community of interconnected beings whose actions pose consequences—or risks—for each other. This of course is in the nature of social life. But is there something special about risks *today* within our technological society that separates our risks and our means of coping with them from those of societies in the past? Is society in the late twentieth century as characterized by modern technology and the risks it imbues more interconnected than societies in the past, so much so that individual risky actions and risky actors are inseparable from each other as part of a new and unique social matrix?

These questions are difficult to answer definitively, since two responses are possible: a qualified "yes," or second, "it does not matter what society is actually like; what matters is how it is perceived." First the yes response. It is at least plausible to argue, I believe, that modern technology spreads the effects of individual behavior more widely than was possible before technologies like television or nuclear fission (to name but two) were available. We *are* more intimately linked to the lives of our neighbors both proximate and remote than in ages past—and modern technologies are directly responsible for shrinking our world. At the very least, we are more aware of the impact of our actions on others and theirs on us, since modern science and technology force us to be so. Still, this awareness is probably only different in degree in comparison to earlier epochs. As Rousseau understood in the eighteenth century, all social life encumbers us in chains, the chains of interdependence that link us to others, the choices they make, and the risks they take. My smoking, sexual habits, reproductive and other choices, or energy usage affect the lives of my twentieth-century fellows, but earlier centuries also had their plagues traceable to the habits of individual hygiene.[18]

In a more significant sense, however, the risks of twentieth-century technological society are different from those of other eras because we perceive them to be so. They are risks not reducible to individuals but emergent in that they seem inescapable and more threatening because of what we know about them. And we do know, simply, *more* about risks than we did a hundred years ago. Partly as a consequence of this increase of knowledge about risk, we also know that no matter what we individually do to protect ourselves, the person next to us puts us at risk by lighting up, the corporation down the street pollutes our air, the government 40 years ago made decisions about nuclear energy that were accepted then but imperil us now. There is no escape from the risks imposed by others, whoever they are.

The anonymity of technological risk both frightens those who live with it and confirms its new, emergent status. These risks are the products of choices or actions made somewhere, sometime else that surround us like a cloud of doubt and fear. And we are at a loss because most of our traditional attitudes toward risk now seem irrelevant. They are. There is no frontier left, no new world, no place where our situation or prospects will be solely the product of our abilities to cope by ourselves. This is a hard realization, made doubly so by the cultural heritage of liberal individualism and its economic counterpart of market capitalism. If, as Harvey Sapolsky says, "Americans live in fear,"[19] it is not only because there are new risks to fear, but because there is no referent in our epistemological legacy that can help us understand emergent risks. These are risks not reducible to their individual parts, so we cannot find someone to blame, punish, or reward. They are not risks about which we have much choice in assuming, as we did when we loaded up the wagons and headed west. These risks emerge from our common life, about which liberalism offers so little insight.[20]

It is possible, then, that the emergent aspect of modern risk is not new but instead a newly realized negation of an important component of our cultural legacy. It is possibly not a feature of a new kind of risk but of a new understanding of what has always been the nature of risk. If it is the latter, the effect on our ability to manage risk is the same. We are without the epistemological tools necessary to come to grips with our risky present. Later chapters explore how individual political concepts within the liberal corpus are affected by the realization of the emergent nature of risk. But in a sense, all that is secondary to the realization itself that risk cannot be reduced beyond the group level. This primary realization requires a change within liberal epistemology about where risks come from and how to recognize them. Coming to this new understanding presumes a huge shift within liberal culture, partly political, partly moral.[21]

Such a conclusion implies at least that the management of risk, maybe even risk itself, is a matter of culture. This point is often forwarded by the school of risk researchers known as culture theorists, who take their cue from early work by Niklas Luhmann, work by Michael Thompson, and a recent book by Mary Douglas and

Aaron Wildavsky entitled *Risk and Culture*.[22] Douglas and Wildavsky summarize their position as follows:

> In risk perception, humans act less as individuals and more as social beings who have internalized social pressures and delegated their decision-making processes to institutions. They manage as well as they do, without knowing the risks they face, by following social rules on what to ignore: institutions are their problem-simplifying devices.[23]

Douglas and Wildavsky present a position that is in some respects similar to mine in that they see the subjective *perception* of risk as an emergent social product. Because it is emergent at the group level, the perception of risk is not merely subjective but intersubjective—that is, it is representative of a whole society. Where we differ is that I believe not only the perception of risk to be emergent but indeed the risk itself. Furthermore, I argue that it is important to keep these two emergent features separate in our understanding of risk. Perceptions of risk are indeed part of risk, but they are not the whole of it. There is an objective, risky reality also, and it too is the emergent product of human group behavior in the development of technology. Culture plays a large role in generating risk behaviors from identification to reaction, but so do science and the objective reality it presumes.

Risk, Rationality, and Culture

The debate between culture theorists of risk and, for lack of a better word, realist theorists is worth exploring, since it illuminates features of my "emergentist" approach that sets it off from both while making clear what I borrow from each. It is a dispute raging since the publication of *Risk and Culture* and one that cuts across several academic disciplines in the sciences, social sciences, and humanities, all of which are relevant to any exploration of risk.

Since Chauncey Starr first asked in 1969, "What is our society willing to pay for safety?"[24] realist risk analysis has claimed a basis in science based on two principles: quantifiability and empiricism. Quantifying risk levels was possible, for Starr, by putting a dollar amount on how much it would cost to eradicate a risk. Based on the amount, it could then be ascertained whether it was cost-effective to do so. In other words, cost/benefit analysis could be applied to the assessment of risk. Second, empirical measures for risk were available that could be studied objectively. Risks were assumed to exist "out there" in the objectively observable world, as did the behaviors of people as they confronted those risks. Thus, the combination of the study of history and observation would reveal the preferences of individuals' concerning risk. The "revealed preference" model of risk assessment was born. Starr summarized his approach as follows:

> I offer an approach for establishing a quantitative measure of benefit relative to cost for an important element in our spectrum of social values—specifically, for accidental deaths arising from technological developments in public use. The analysis is based on two assumptions. The first is that historical national accident records are adequate for revealing consistent patterns of fatalities in the public use of technology. . . . The second assumption is that such historically revealed social pref-

> erences and costs are sufficiently enduring to permit their use for predictive purposes.[25]

Starr assumed that risk policy could be generated based upon information derived from the revealed preference model that could then be plugged into an equation of monetary costs and benefits. If studies showed that citizens lived unquestioningly with a certain level of risk, risks less than that level could logically be assumed to be acceptable to those citizens. The goal of such an exercise was rationally to plan for the development of risky technologies without causing alarm among the citizenry. Risks could be studied and anticipated in a cool, logical, scientific manner, employing models of decision based on rational calculation.

Unfortunately for the revealed preference approach, later psychological studies called into question whether there was any logic to the behavior of people in risk situations. The work of Paul Slovic and Baruch Fischhoff, among others, provided ample evidence that the economic calculations of the revealed preference model presumed a degree of realism about dangers and a transitivity among preferences that were rarely present in the way real people thought about risk. Cognitive psychological studies repeatedly showed the same surprising errors in the way everyday citizens assessed risks. For instance, people frequently overestimate the risks of dramatic methods of harm or death, such as airline disasters. Furthermore, outside elements such as media coverage or popular recent movies affect the feeling of dread people experience in the face of risky events.[26] Also, people tend to exaggerate the risks of catastrophic risks generally, as compared to chronic ones. They fear unfamiliar threats more than ones they know, downplay risks to which they feel they have consented, and frequently misunderstand what probabilities mean when assigned to risks.[27]

The barrage of psychological and social science studies carried out throughout the 1970s and 80s seriously challenged the idea that risks existed solely "out there" in the external environment and could therefore be confronted in the logical, rational way that Starr had presumed. These studies showed clearly that the way people felt about risks had as much to do with their psychological state at the time of confrontation with risk, with values they held about seemingly unrelated topics, or with attitudes they held about policy makers in both the public and private sector involved in risk decisions, than with the observable level of risk ascertained through empirical methods.

Critics of Starr's approach also pointed to prejudices built into his model that were the result not of scientific objectivity but of systematic bias. For instance, an underlying assumption of the revealed preference model is that society has achieved "an essentially optimal balance between risks and benefits"; thus, new risks could be evaluated in terms of whether they exceed the level of risk of ongoing technologies or practices.[28] However, not all citizens might agree that the current level of risk or its distribution is optimal. Those who do not benefit from the current regime might well object. To take an example we explore more fully later, the poor generally run worse risks than the wealthy, as Robert Goodin argues; thus, the revealed preference model—and related economic approaches—is an ideological justification of and excuse for the current maldistribution of risk in technological societies.[29]

Proof of the extent to which the inequitable distribution of risk in society is now finally being recognized can be seen in President Bill Clinton's 1994 executive order requiring all federal agencies to develop strategies for "identifying and addressing disproportionately high and adverse human health or environmental effects of [their] programs, policies and activities on minority populations and low-income populations."[30] Concerns for "environmental justice" now encompass many areas of risk exposure and have spawned new studies that explore possible economic class and race-based inequities overlooked since the revealed preference model first took hold as a decision strategy for risk policy. Siting disputes of all kinds now frequently raise charges of environmental racism or economic injustice, but other examples include allegedly differential exposure to pollution generally and the specific case of exposure to the risks of lead poisoning. The latter example has generated much study and literature since the late 1980s.[31]

Finally, the belief in science and its alleged neutral objectivity may itself be a prejudice not shared by all who experience risk. Much literature in the philosophy of science calls into question the whole idea of value-free endeavor as the hallmark of the scientific enterprise. Thomas Kuhn first raised the specter of political ideology and personality as the real historic motor of scientific revolution. Also, many feminist philosophers of science hold up the elements of the scientific creed, including presumptions about the universality of reason, as just so much cultural baggage within the long history of patriarchy in Western societies.[32]

I further explore these critiques of both science and risk measurement techniques in the next chapter, as we view their impact on the liberal understanding of consent. Here what is significant is that from these critiques sprang the cultural approach of Thompson, Wildavsky, Douglas, and others. Beginning with an assumption that attitudes about risk were the product of culture and not the "real" external world perceived "rationally," they concluded that risk itself is a collective product that varies widely across cultures. Little about risk is absolutely or universally true, they concluded, since both the attributes of risk and attitudes toward it are relative to culture.

The culture theorists clearly build upon the studies of the cognitive psychological approach, in that they seek to understand risks in terms of how people perceive them, not in terms of risks' allegedly objective reality. Indeed, for some culture theorists, that reality either cannot be measured or is irrelevant to the question at hand, though the second eventuality tends to follow in positivistic fashion from the first. Sheldon Krimsky, for example, concludes that "in studies of risk perception, in many instances there is no objective measure of risk."[33] The cultural approach broadens the perceptions from individuals to society as a whole, thus stressing an "emergent," collective subjectivity. But the two approaches share a basic belief that risk is as much a matter of values as it is a matter of facts or reason.

The cultural approach, as it is presented by Douglas and Wildavsky and others, has been criticized for its own biases toward industry[34] and political elitism.[35] Nevertheless, it offers a powerful critique of earlier rationalistic approaches toward risk. Furthermore, the relativism implicit in culture theory fits well with the "postmodern" criticism of modernity's catechism of scientific and ethical absolutes, as that criticism is offered by diverse thinkers including feminist theorists, Michel Foucault,

Jacques Derrida, and others. Thus, culture theory ties attitudes about risk to larger aspects of the contemporary mind in an effort to understand the connections between technological risk and a host of societal institutions and group behavior.

This refusal to reduce analysis of risk to that of individual behavior is one of the strongest arguments that culture theory makes in its approach to risk. Less successfully however, this theory also denies the validity of other epistemological reductions or assumptions in our understanding of risk. It contends, for instance, that there is no incontrovertible objective physical reality about risk apart from its (culturally defined) perception. At the very least, some culture theorists maintain, knowledge of the physical side is not always very useful. Michael Thompson says, "It is the social, not the physical, understanding that is likely to lead us to more effective ways of handling technological risk, especially in those instances where, try as we may, we simply cannot gain the physical understanding we desire."[36]

Culture theory also denies that absolute standards of rationality can be invoked to assess the response to risk. There is no single definition of what counts as rational choice or behavior. Rather, like political institutions, perceptions, or values, "conceptions of rationality come to be selected in ways that best preserve a particular social context."[37] This denial of the universality of reason is perhaps refreshing when compared to the overly simplistic models of rationality emanating from some rational choice theories that presume that eighteenth-century models of economic rationality are absolute in all times and places. But culture theorists can sometimes stumble over their own relativism. Shrader-Frechette accuses cultural relativists of arguing that, "because everybody can be wrong at some time, everybody is wrong all the time."[38] This is a bit unfair on Shrader-Frechette's part, because "wrong" is also a relative concept in cultural context. Nevertheless, it is warranted to accuse Douglas and Wildavsky, as Shrader-Frechette does, of their own absolutist prejudices in their subtle attribution of irrationality to positions other than their own. *Risk and Culture* presents a palpable bias toward industry and does indeed "single out US environmentalist or sectarian laypersons (as opposed to technical experts) as having particularly biased constructs."[39]

Ultimately, the dispute between culture theorists and realists about the source and nature of risk is, as Shrader-Frechette asserts, "a conflict over values." It is not my intention to mediate this debate but to make three points about how viewing risk as an emergent phenomenon can bring these two approaches together. This theoretical rapprochement is particularly enlightening if the second salient—but underappreciated aspect of risk is also given due credit: the essentially political nature of technological risks.

First, the cultural approach performs the invaluable service of pointing out that the subjective side of risk is a matter of collective culture, not merely individual psychology. Few scholars any longer deny that the ontology of risks manifests both subjective and objective features. But it is important to realize, as the culture theorists make clear, that the subjective construct of risk is itself clearly an emergent one, the product not of only one person's experience, but of a society's past, present, and anticipated future. This is true whether individuals recognize it or not, since the effects of culture are not always consciously acknowledged by affected individuals. Still, it is certainly true that some cultures produce a better understanding of this aspect of risk. For instance, the liberal individualistic past of US culture makes it notably difficult for us to see this point about the emergent nature of risk, to our considerable discredit as well as discomfort, as later chapters will make clear.

Second, in her partial defense of the realists, Shrader-Frechette is also correct in maintaining a role for a realist view of science that presumes an objective external world that presents "real" risks. Her approach is a compromise:

> But even though risk evaluation is not wholly objective, neither is it *merely* evaluative nor only a construct. Constructs don't kill people; faulty reactors, improperly stored toxics, and poor risk evaluations do. At least some hazards are real, and many of these are measurable. Hence, the cultural relativists are wrong in overemphasizing value judgments in risk assessment and evaluation. Risk evaluations assess *real hazards*, not just constructs.[40] (first emphasis Shrader-Frechette's)

Shrader-Frechette's compromise is somewhat close to culture theorist Niklas Luhmann's distinction between dangers and risks. Hazards do exist as dangers whether recognized or not, according to Luhmann, but it is only fruitful (at least for culture theorists) to refer to them as risks when attitudes coalesce around them that bring them to the public consciousness.[41] Similarly, Sheila Jasanoff tries to bridge what she somewhat disingenuously names the two "cultures" of risk analysis by correctly recognizing that technological artifacts that are often the causes of risk are themselves "continually constructed and reconstructed through social practices. Risk is the product of this constant interaction; it is not simply there to be 'found' as an unalterable characteristic of an inanimate physical system."[42]

Jasanoff's proposed synthesis recognizes that Shrader-Frechette's "real hazards" *are* indeed constructs in at least one very important sense. They are the product of group decisions and group activities within a society's past or present technological development. Just as important, those pasts and presents are different in different societies. This cultural variability may seem obvious and trivial, but it is important to realize, for example, that it is a very real *cultural* as well as physical difference in the world today that some nations are nuclear powers (militarily and in civilian power generation), and others are not. Thus, nuclear energy is a (dangerous) fact both about the "real" world and about cultural difference. That fact (present or absent in different societies) is the product of a long history of scientific discovery, technological and economic choice, and policy making within different societies—it is a product of a society as a whole. And whether a given society possesses nuclear capability says something both about its culture and how it differs from other societies.

It cannot, incidentally, be accidental that societies with liberal pasts were first to embrace nuclear technology and its attendant risks and remain societies most likely to rely on nuclear generation of energy. The liberal attitude toward science, technological change, and the risks generated thereby was crucial in developing nuclear as well as other modern technologies. The liberal understanding of the adversarial relationship of human beings to their environment is an important element of liberal culture that fed the development of much of modern science and its various implementations such as nuclear technology. It is an emergent fact about liberal culture to which we will return in later chapters.

Thus, the second way in which the emergent view of risk brings the culture theorists and realists together is by recognizing that differences of technological development and risk in different societies are objective facts. That is, some societies possess technologies and risks absent in others. But also, those differences are themselves the emergent products of varying cultures manifesting different attitudes toward modern science, technology, and their attendant risks. Therefore, these differences are emergent objective *and* cultural facts.

They are also political facts. Shrader-Frechette's "real hazards" are constructs in another sense as well. They are the product of a political reality, of political choices about allocation of money for research and development, about distribution of risks, and about the exercise of power. This is the third essential feature of seeing risk as an emergent phenomenon. So far we have seen that risks are both "real" and culturally defined and that they emerge at the group level. Now we must explore how that level is more clearly describable in political terms than in any other. In so doing, we will see that the politics of risk is more than merely policy battles among bureaucrats or political leaders. It involves workers, employers, doctors, patients, parents, children, *all* citizens in a variety of their everyday and professional roles. The politics of risk is not entirely contained within the definitions of constitutions or the descriptions of governmental institutions. It is a politics that is pervasive in all aspects of life—it emerges, and in the recognition of its emergence lie both the true understanding of modern risk and the synthesis of the realist/culture theorist debate about risk ontology.

Risk and Politics

Neither culture theory nor realism recognizes sufficiently the political side of modern technological risk. Each of them, as well as mixed theories such as Shrader-Frechette's and Jasanoff's, surely leaves room for a political look at risk, and certainly most of them would not deny that risk can assume political dimensions.[43] But it is not enough to say that risk is frequently something about which political decisions are made, or even that today virtually all political decisions and issues have within them some component of risk emanating from a relevant technology.[44] What I want to argue is not that politics is something *added on* to risks after they emerge either from nature or social interaction, but that at their core risks are essentially political phenomena. Doing so will also render moot the entire debate over their identity as either technical (objective or "real") or cultural phenomena.

A growing literature in the field of risk studies recognizes the increasingly important political aspects of risk, without necessarily concluding, as I will, that risks are most accurately described qua political things. Beck probably goes the furthest in seeing the essential politicalness of risk, since his contention that power is both pervasive in the "risk society," and is exercised almost exclusively for the distribution of risks rather of wealth or benefits, is surely at bottom a political thesis. For instance, he says:

> In contrast to all earlier epochs (including industrial society), the risk society is characterized essentially by a *lack*: the impossibility of an *external* attribution of hazards. In other words, risks depend on *decisions*; they are industrially produced

> and in this sense *politically reflexive*. While all earlier cultures and phases of social development confronted threats in various ways, society today is *confronted by itself* through its dealings with risks.[45] (emphases Beck's)

Notwithstanding his obviously political interpretation of risks, Beck overstates his case by insisting that an external attribution of hazards is "impossible." Seeing risks as emergent means partly that attributing the causes or sources of risk is certainly more difficult than we once thought, but only because we need to get away from an excessively reductionist ontology of social phenomena. Beck is saying something different, however, because he assumes he is exposing the inescapable dynamic of the modern risk/power/knowledge nexus. In so doing, Beck excessively reifies the whole phenomenon of modern risk by characterizing it as a product of society that cannot be avoided or even fully recognized as dangerous, since in viewing risk modernity cannot see past its own "reflexively" projected self-image. Although here his view of risk as emergent is unmistakable, Beck renders the whole enterprise of coping with risk far too fatalistic, if not itself politically impossible. As I argue, the full recognition of risk's reality does indeed expose liberal society's flawed self-image as excessively reductionistic—liberalism is indeed "confronted by itself" because of risk. But in that recognition lies the realization that a different conception of society, at present "external" to liberalism's self-definition, is indeed possible, one that if embraced also encompasses the political means necessary for coping with risk. I am, in short, more optimistic than Beck about what is possible once we attain a fuller understanding of risk's underlying nature.

In addition to Beck, several other risk scholars have highlighted different aspects of risk's political character in a way more illustrative of that character. Not surprisingly, many of them adopt a culture-theoretic approach to risk. For instance, Ingar Palmlund argues that whenever a society "deals with risks, whether of aggression, natural hazards, or a controversial technology, it is engaged in a political process." That process is marked by the necessity of choice between various tangible and symbolic goods and costs, requiring decisions not only about quality of life and economic values, but about the social definition of reality itself.[46]

Also within the culture theory camp is Brian Wynne, who has focused attention in many articles and books on the role of political authority in the definition of risks. For Wynne, the essential question about the presence of risk is usually not one of technical measurement but of the measure of trust in institutions.[47] To the extent that today some citizens seem to be calling for a life free of risk, Wynne suggests that such "'zero risk' demands should perhaps be interpreted as expressions of 'zero trust.'"[48]

Wynne's approach extends beyond the bounds of culture theory, since in his focus on the power to define risk, he invokes both political and psychological aspects of risk definition. In the first, he is echoed by William Leiss and Christina Chociolko, who insist that most risks presume the exercise of power by someone else:

> First, risk means "exposure to the chance of loss"; and second, for many (but not all) of the risks that we in modern society encounter, our exposure is deliberately induced by some social actor in order to realize an incremental net benefit. None

of the disagreements about risk that pervade our society are comprehensible unless we see the necessary connection between these two propositions.[49]

Even outside the culture theory approach, the role of politics is highlighted by risk analysts adopting a more psychological model, thereby supporting Wynne's political approach in a second sense. In the newly developing subfield of risk communication studies, the recognition of risks is held to be largely a result of how risks themselves are communicated to those experiencing them. That process is fraught with political manipulation and the exercise of power. Sheila Jasanoff contends, for example, that the very term "risk communication" is often a code word for "brainwashing by experts or industry."[50] Unlike Jasanoff, Morgan et al. do not see risk communication as an intrinsically nefarious enterprise; nevertheless, they acknowledge that it necessarily invokes the power behind "the choices people face, what beliefs they hold, and what expert knowledge exists."[51] It is that power that is most evident in a vision of risks as distinctively emergent phenomena.

Seeing technological risks as intrinsically the products (if not the whole) of politics is not difficult once we accept the emergent nature of those risks. If risks emerge from the social processes involved in scientific discovery and technological development as those processes are manifested in culture, then it is those processes that must be political if what they produce—risks—are political phenomena. It is surely not difficult to see that at least *some* processes of discovery and development are political. Some interpretations of Kuhn's famous study of scientific revolutions contend that his theory requires the conclusion that discovery is always a matter of politics.[52] A paradigmatic example of development being a matter of politics is, of course, the Manhattan Project that produced the atomic bomb and began nuclear technologies generally. Furthermore, modern technologies that carry risk are frequently the focus of political debate and the object of political decision at various levels of government and politics. Still, this by itself does not show that *all* technological risks are (intrinsically) political phenomena.

What does confirm the essentially political nature of risk is the fact that risk and politics share the same two defining features. Politics is subject to a host of definitions of course, but at least since the time of Machiavelli and his US interpreters—the authors of Federalist Papers—politics is most often defined as the interplay of two basic elements: conflict of interest and power.[53] What I want to demonstrate is that technological risk is the product of the same two forces. Of course, we would not be surprised to discover, I think, that risk can also be the cause of conflict of interest and the increase (or decrease) of power. The same can be said for politics.

Technological risk is the product of the conflict among interests held both by the human species as a whole and by its individual members (or their relevant groupings). In an evolutionary sense, the development of technology is a characteristic of our species, and an almost unique one. Humans devise and use tools. We do so in the interest of survival, whether that survival is described in terms of our genes, our species, or our various cultures.[54] Our survival is our own interest, of course, and often at odds with the interests of members of other species (sometimes whole species), given the scarcity of resources. The fact that our technologies can in fact threaten our own species' survival as well is ironic to say the least but does not alter the fact that our ability to develop them is an adaptive response to further our evolutionary interests. That human technologies also threaten other species' survival is partly a matter of the evolutionary principle of survival of the fittest and partly of our special ability as a species to wreak havoc with the natural order. In any case, the risks our technologies carry are clearly products of our species' interest in survival, an interest often at odds with interests of other species.

On a less cosmic scale, technological risks are the product of the conflict of interests among individuals and groups within societies, sometimes within different societies, and also between societies taken as wholes. Again, the relevant vehicle of competition is scarcity, usually defined in terms of wealth, but strictly speaking, it is a scarcity of resources. Technological development in modern societies is usually the result of governmental action or of investment in the marketplace. That is to say, modern technological innovation is a matter of public policy or private enterprise, though the distinction between the two is increasingly murky. In any case, what is at issue is either the capture or redistribution of resources at the level of government, or the competition for wealth among individuals (either corporately or alone). Thus, the risks that entail from the development of technologies emerge, again, from the conflict of either political or economic interests.

Both the pursuit and clash of interests presume the presence of power, almost certainly in differential measure among the actors involved. It is a commonplace that technological development requires power of several types: economic, intellectual, even electrical. Power is exercised in several locations in the process of development as well: in government institutions, both electoral and administrative, in industries, in laboratories, in the workplace. The power that gives rise to technological risk is of several types then: political in the sense of office, economic, intellectual, administrative. We have much to explore concerning the nature of this power, but first two preliminary points.

First, the power to create new technologies, and therefore to impose the risks they entail, is an ability to affect other people, thereby making it a political power in the traditional interpretation. The nature and extent of these effects is not always immediately known, as, for instance, in nuclear or reproductive technologies, nor do we always know how these risks affect our environment as a whole. But it is indisputable that the exercise of technological power puts others at risk. Furthermore, if one wishes to make a distinction between the exercise of power and merely "having" power, as some philosophers are wont to do, it should be pointed out that many technological innovations dissolve the difference.[55] For example, if all nuclear weapons were destroyed tomorrow, we would still possess the knowledge and ability to recreate their ominous threat.

This tendency of technological power to overwhelm the distinction between its exercise and merely its possession demonstrates a second important aspect of this power. It is a power with, in the words of Bachrach and Baratz's well-known formulation, "two faces."[56] All political power, according to Bachrach and Baratz, has two sides or faces, the power to make things happen and the power to prevent them from occurring. The second is more subtle because it often is not seen as power at all. In a political context this power is most often in use when issues are *not* placed on the agenda of government or of social discourse—in other words, when they are not recognized as issues. Power is in use here, they maintain, but it is a prior power exercised in the prior "mobilization of bias" in the community. It is a power manifested in "the dominant values and the political myths, rituals, and institutions which tend to favor the vested interests of one or more groups, relative to others."[57]

The power emanating from technological risk also functions in this second mode and, because it does, should be seen in this fully political light. We live with many risks today that are the by-products of technologies; in many cases we do so unquestioningly. We have become so inured to (and biased in favor of) the ceaseless march of innovation in technological fields such as communications technology, to take but one example, that the risks (e.g., to privacy) ahead on the "information highway" or along the communication satellite's orbit are not recognized until they are upon us, if even then.

We need to explore the natures of power and conflict within technological politics further, since, as we will see in later chapters, they threaten to overwhelm the centuries-old concepts that have defined our liberal approach to political life. In the remainder of this chapter I introduce three themes that will inform our exploration of the changing conceptual basis of liberal politics that technological risk has effected. First, in its basis in power and conflict of interest, technological risk has first affected the scope of politics, enlarged it to the extreme of pervasiveness. Second, risk affects the democratic character of politics, heightening democracy's strength in some ways, diminishing it in others. Third, modern risk exposes the flaws and failings of liberalism as a basis for either democratic or technological politics.

The Pervasive Politics of Technological Risk

Today technology's risks are all around us, and we know it. We live with them at home in front of our televisions and computer screens. They invade our bedrooms in our sleep and in our lovemaking; they reside in the appliances in our kitchens and in the air conditioners in the window. We travel with them to work, and they await us when we arrive. They accompany us over the weekend and on vacation. They are in the water, the air, the old asbestos walls, the food, the food chain. They pervade our lives and invade our sensibilities. If we pay attention, we know we are living on the edge.

It is not a revelation that modern life is full of risks and modern science full of warnings. What is new is the realization that the dread we experience is a recognition of the power that risk (and knowledge of risk) exerts in our lives. As I argued earlier, this power needs to be seen as a feature of the conflict of interests that produces emergent technological risks. Simply put, the risks to my health or livelihood are the risks someone else generates in the name of profit, self, or country. In other words, the power of risk is a political power that engages us all to some degree. And because we experience that power in virtually every aspect of our social lives, the politics of risk, we must also conclude, is everywhere also. Like risk itself, it pervades our lives.

Our traditional liberal understanding of the scope of politics bridles at this insistence that risk renders every aspect of our life political. Since Hobbes wrote *Leviathan*, it has been at the heart of the liberal faith that politics can be segregated into a public arena that we enter only as a matter of choice. Privacy is the birthright of the American revolution in the minds of US citizens; none of our precious rights and liberties can exist without it. Politics under this liberal formulation is a regrettable necessity meant to preserve the private realm, not to invade and overwhelm it. Even though not articulated expressly in the Constitution, the right to privacy is the aura, the penumbra surrounding and illuminating all other rights. We will see in chapter four that, perhaps more than any other, the right of privacy has been redefined by modern risks.

It is not immediately clear how modern technological risk threatens privacy. In fact, many contemporary technological advances seem to enhance the private, nonpolitical kingdom of every citizen. Television, telephone, and the Internet allow us to see the world and communicate with whom we wish from the private anonymity of home. Furthermore, as will become evident in chapter four, it was advances in reproductive technology—specifically in birth control—that first gave legal recognition to the right of privacy by instigating the case of *Griswold v. Connecticut* that led to the Supreme Court's famous "penumbral" doctrine.[58] But these technologies politicize the privacy we enjoy specifically by bringing the power of government—and of others—into our lives. Telephones make private conversations possible with distant friends and allow "telemarketers" into our dinner hour. Television lets us stay in and be entertained and opens our living room to corporate America. The Internet opens communications with the world and raises the possibility of government encryption of computer messages.

Whether we agree or not that such technological developments represent palpable risks, it is clear that they engage us in the lives of others—and they in ours—in new and still evolving ways. These interactions do not escape the taint of power, and thus they are political. It is crucial to individual liberty that we see these interactions as political, or else we run the risk of being enslaved by elements of power over which we never even try to gain control. Many political theorists have argued this point—that if we really desire liberty we must jettison the blind faith of liberalism that instructs that only government exercises political power and thus must be subject to democratic controls. Rousseau and de Tocqueville, of course, never limited the "chains of modernity" to those forged by government. Similarly, Carole Pateman argues that it is central to an adequate theory of democracy that "individuals and their institutions cannot be considered in isolation from one another," and the "political" not be "confined to the usual national or local government sphere."[59]

Many participatory and deliberative democracy theorists follow Pateman in her insistence that the old, liberal rendering of politics as a narrow realm of governmental activity must be jettisoned if either democracy or liberty is to be maintained.[60] Politics understood as the exercise of power and the conflict of interests permeates all aspects of our lives. Not all those who understand politics' ubiquity, however, are as optimistic about democracy's adaptive ability to respond. Jurgen Habermas sees the power of politics as a domination over the individual that only a wholesale reconstruction of civil society can remedy.[61] Michel Foucault envisions an even more difficult task ahead, since "modern man is an animal whose politics places his existence as a living being in question."[62]

Foucault identifies what he names "bio-power": a convergence of political power as exercised by the state with the modern technologies capable of creating, altering, and sustaining human life. When the state comes (as it has, he claims) to see the general growth and care of the population as its rightful mandate, then the regime of bio-power has begun. Bio-power brings "life and its mechanisms into the realm of explicit calculations and makes knowledge-power an agent of the transformation of life."[63] Such is the way of modern science, according to Foucault, especially as it has become the primary tool of state power. But the knowledge to transform life is not the sole property of the state in Foucault's view. In "free" societies that knowledge is disseminated through communication technologies to all comers. Thus, bio-power becomes the property of industry, media, medicine, education, and all institutions in society, creating a power that transcends mere governmental structures based on law. It is a power that defines acceptable patterns of behavior not by relying solely on government-enforced, official threats to punish those who transcend the bounds, but by evoking a subtle internalization of control within the individual himself or herself who becomes "normalized":

> Another consequence of this development of bio-power was the growing importance assumed by the action of the norm, at the expense of the juridical system of law. . . . The law always refers to the sword. But a power whose task is to take charge of life needs continuous regulatory and corrective mechanisms. . . . Such a power has to qualify, measure, appraise, and hierarchize, rather than display itself in its murderous splendor; it does not have to draw the line that separates the enemies of the sovereign from his obedient subjects; it effects distributions around the norm. . . . A normalizing society is the historical outcome of a technology of power centered on life.[64]

The effect of normalization as the consequence of power is somewhat similar to Bachrach and Baratz's understanding of power as the "mobilization of bias" in society. Its significance here is in how Foucault locates the technological aspect of power as pervasively dispersed, or, in his terminology, "capillary," throughout all of society, thereby making it difficult to recognize its application, much less be able to resist it. Ultimately, as Foucault's famous metaphor of the prisoner within the panopticon is meant to show, the coerciveness of bio-power is internalized by every individual within society, normalizing him or her to "watch" himself/herself, making external surveillance superfluous. This is the final eradication of human life in the sense of free individuality—politics as bio-power is completely internalized by every person, making all of life subject to norms that emerge from the nexus of technological growth and scientific knowledge.

Obviously, Foucault's discouraging vision of life in technological society is his own and not necessarily shared even by others who with him deny the narrow liberal definition of the scope of the political realm. But what these authors share is a recognition that the effects of modern technology bring individuals together in a way that dramatically affects the public realm as well as the private sphere. We can no longer view with any accuracy politics as something that exists "out there," somewhere away from our living rooms. Politics imposes itself into our lives at every moment, since technology and its risks have brought us together into a public realm from which we cannot depart.

The Democratic Politics of Technological Risk

All this togetherness might seem to make technological risk appear as a democratizing force in modern society. Indeed, Theodore Lowi discusses what he calls the "democratization of the costs of risk taking" as part of a whole "new universe of ethics" imposed by modern risks.[65] But we should be careful before ascribing too much democratic spirit to the new environment of risk in which we find ourselves. Modern risks do affect politics in ways that affect democracy; much of the rest of this book is an exploration of that impact. Some of risk's effects do seem democratic in some sense, but mostly the relationship between risk and democracy is ambiguous. Modern risk opens the door to new experiments in democratic decision making, as we will see in chapter six, but also lays new impediments in the path of democratic development.

On one hand, its emergent character enhances modern risk's democratic appearance. Risks emanating from air pollution or radioactive waste, for instance, affect everyone regardless of their level of political influence—no one can fully escape them. Similarly, as Lowi points out, making policy for emergent risks involves the resources of society as a whole; thus, costs are "democratized" in a sense. On the other hand, this seemingly populist effect is easily exaggerated since costs are rarely distributed equally. We all live with pollution, for example, but some of us can afford homes at the seaside or in the country where the air is less noxious. Similarly, radioactive waste is a problem for us all, but, of course, living next to a disposal site has obviously inequitable effects for some unfortunate few.[66]

Siting disputes are made of such risk inequities, and as a consequence Lowi sees a more significantly democratic effect of coping with modern risk. Modern risks, he believes, invoke the language of democracy in ways new to our day. "Recent policies reallocating risk and the responsibilities and costs of risks," Lowi argues, "have been framed in such a way as to convey rights to the beneficiaries."[67] "Rights talk" is an essential feature of democratic politics and has been since we held certain truths to be self-evident in 1776. But the definitions of what counts as a right and of the obligations it imposes on others are both evolving within disputes about modern risks. These disputes will have important effects on the meaning of democratic terms such as rights and responsibilities and, therefore, on the nature of democracy itself; I will explore them fully in chapters four and five.

Two other elements of the interplay of risk and democracy are worth noting here by way of introduction and to which we will return in chapter six. First, modern risks may be changing the meaning of democratic participation. Though voting remains at a relatively low level in the United States when compared to other democracies, other forms of political activity are increasing in importance as a gauge of overall participation. Chief among them are grassroots organizing efforts to promote or hinder policy making for risk. These efforts are most often evident during siting disputes but have spilled over into areas of general, nationwide policy formation. For example, many *localities* have experienced the tumultuous politics of nuclear energy either in terms of the siting of reactors or of waste facilities.[68] But it is unquestionable that the nuclear fears raised by such local disputes fueled the *national* consideration of the nuclear freeze movement of the early 1970s.[69] Technological risks have, in short, increased the level of popular participation in politics throughout the past three decades.[70]

Second, it is a motivating thesis of this book that the emergent nature of technological risk compels us to make risk decisions in a more participatory manner. As a believer in deliberative democracy, I believe this to be true partly on ethical grounds—it is, I believe, the right thing to do. But the empirical grounds are persuasive also. Disputes over radioactive waste disposal (among other technological risks) have taught us that it is simply not possible to arrive at policy without sincerely democratic efforts being made to maximize participation. Citizens have become very adept at blocking policy by using media, the law and the legal profession, and the technological fears of their fellow citizens. Whether we applaud or regret these efforts, they have raised the ante in the policy struggle between experts, lay citizens, and organs of government. If policy is to be made for the emergent technological risks we face—and our survival demands that we make it—new models of democratic participation must be explored. We will do so in chapter six.

Conclusion: The Illiberal Politics of Technological Risk

The deftness on the part of citizens in blocking policy regarding certain risks, as in siting controversies, illuminates an additional aspect of the politics of modern risk. Risk politics often undermines in a number of ways the liberal heritage shared by most modern technological democracies. We have already seen that the emergent nature of risk befuddles the liberal epistemological reduction of all phenomena to their individual components, even to the extent of making the concept of agency unintelligible in totally individualistic terms. We have also noted how the liberal idea of individual liberty is threatened by the "normalizing" power of risk politics. The next chapter explores how risk challenges the notion of consent that remains so central to liberalism's understanding of political obligation and legitimacy.

In the context of political participation, too, risk challenges liberal definitions. Fundamental to the liberal idea of government are three faiths: in reason, in representative institutions, and in the sanctity of rights. Both philosophical and pragmatic approaches to risk politics challenge all three. Foucault again sounds the philosophical challenge. Paul Rabinow points out that Foucault "refuses to see reason as either our hope or our nemesis." Rather, in Foucault's words, "[t]he relationship between rationalization and excesses of political power is evident. And we should not need to wait for bureaucracy or concentration camps to recognize the existence of such relations."[71] Since large portions of the existent government bureaucracy have been created specifically to cope with modern risks, the persistence of those risks in unameliorated form is a warning about the unreliability of reason when applied to governmental administration.

In the practical politics of risk policy the loss of faith in representative government is a palpable part of almost every siting dispute.[72] Chapter three looks closely at the ideas of authority at work in these disputes; suffice it to say here that citizens do not trust their elected representatives to protect their interests when threatened by risk. Whether the authority invoked comes instead from the qualifications of technical "experts," or from the moral claim that living with the consequences grants to all citizens the authority to decide, what is clear is that, as a consequence of modern risk, elective office is no guarantee of authority in the eyes of an increasing number of citizens.

Furthermore, in these disputes the authority that once resided in the claim of "rights" is also severely challenged. Technological society simply cannot abide all the claims of NIMBY that proceed on the basis of (allegedly) inalienable rights. It (whatever) must inevitably be in *someone's* backyard, and the individual's right to deny society's claim in effect denies the rights of many other persons, living now or in the future. Interestingly, as we will see in chapters four and five, consequentialist moral theories both sustain the argument against elective authority and deny the argument supporting the authority of rights. Liberal moral theory will go through severe contortions before disputes over risk subside, as will liberal politics. The emergent nature of modern technological risks is a challenge to most, if not all, liberal concepts of politics and morality. It is to the elaboration of both that challenge and liberalism's potential response that the remaining chapters are devoted.

Liberalism contains within it many epistemological, ontological, social, and ethical positions and presumptions. But fundamentally it is a doctrine of politics, one that begins with the premise that the realm of politics should shrink to the level containable within the "minimal" or "night watchman" state. As a political theory, then, liberalism embraces concepts meant to forward the goal of a shrinking (and shrinkable) state. The next chapters consider the most important of those conceptualizations of liberal theory in an attempt to show two things about politics and liberalism. First, even though (or perhaps because) the heritage of modern science and technology is ineluctably tied to liberalism as a fact about both of their historical growths and developments, modern risks as the products of this marriage now manifest the need for a divorce. Second, the language of risks and the making of policy for them is now fully suffused with the language of politics. Thus, to examine the reality of risks today we need to consider the concepts of consent, authority, rights, responsibility, and participation. These are recognizable as the building blocks of liberal theory both as practiced beginning in the seventeenth century and continuing today. But their meanings have been so drastically altered by the risky progeny of the marriage of liberalism and science that their traditional liberal meanings now either seem irrelevant or at least detrimental to effective policy making. The politics of risk has adopted the language of liberalism but in so doing has proven that language's inadequacy.

2

Risk, Consent, and Communal Identity

No belief is more central to the liberal impulse in politics than the value and legitimating power of consent. Our beloved prepositions "of, by, and for" the people characterize a government rooted in the sacred ground of voluntary acceptance by its citizens. Individual liberty begins for liberal theorists from Locke onward with the right of every person to voluntarily assume political obligations. Today most liberal theorists and political scientists agree that only the "consent of the governed" can legitimate the power to enforce obedience and to recognize political obligation, even though, as Richard Flathman once wryly observed, accepting most consent theories of political obligation quickly leads to "the conclusion that very few people have or have ever had political obligations."

Consent is also an important feature of technological risks and people's attitudes toward them. In an early study of the "revealed preferences" constituting the public's attitude toward risk, Chauncey Starr wrote that consent was such an important feature of risk attitudes that "the public is willing to accept 'voluntary' risks roughly 1,000 times greater than 'involuntary' risks."[1] Similarly, culture theorists Mary Douglas and Aaron Wildavsky include consent in the very definition of risk as a "joint product of *knowledge* about the future and *consent* about the most desired prospects."[2] Further, Baruch Fishchoff and his colleagues explore several models of risk assessment to show that all of them make assumptions about the nature and presence of consent in risk decisions. They assert that the issue of consent is furthermore a distinctly political brand of consent, since choosing an assessment model "is a political act that carries a distinct message about who should rule and what should matter."[3]

As we have seen in chapter one that technological risk decisions dominate politics today, it should come as no surprise that issues of consent present in risk assessment decisions are pertinent to the larger question of consent that lies at the foundation of liberal society. This chapter and the next explore aspects of that relationship. Today there are instances in every morning newspaper of citizens rescinding or refusing to grant their consent to risky technological development projects ranging from the siting of hazardous installations to recombinant gene research and therapy. In so doing, they contest the tacit consent arguments that, as we will see in this chapter, lie at the heart of the revealed preference and most other models of risk decision making.

Many controversies involving NIMBY claims particularly single out tacit consent as the objectionable feature of siting decisions.[4] I explore the nature of tacit consent in this and the next chapter but focus especially on how it operates in two different contexts within liberal society and as part of risk policy controversies. Most risk issues call attention to one or both types of consenting on the part of citizens, whether express or tacit. First, risk disputes invariably involve issues of consent to persons claiming sufficient and legitimate authority actually to make policy decisions or to inform them. Authority in such cases can be based either on elective office, administrative position, or technical expertise but is grounded in consent that takes the form of citizens' acceptance of decisions made by the person in authority. Thus, consent takes the form of citizens' deference to those persons who are properly constituted authorities and to the decisions they make. For instance, for radioactive waste disposal policy, authority may reside in the hands of a governor, legislature, agents of the Nuclear Regulatory Commission or Department of Energy, or in the minds and experience of nuclear experts such as physicists, geologists, or other scientific experts.

As we will see in the next chapter, some risk controversies are explicable only by recognizing the degree to which all or some of these bases of legitimate authority are being rejected by citizens in their opposition to policy. But, in any event, consent here needs to be seen in its first manifestation as deference to particular persons or positions charged with decision-making or decision-informing authority. Relevant issues in assessing the validity or bindingness of such consent include, first, whether the consent is truly "informed consent" and what that qualifier means. Informed consent as a concept is one of the products of the decisions at Nuremberg following World War II and has been progressively adumbrated in a series of US federal court cases as well as many scholarly treatises.[5]

A second issue involved in consenting to the authority of persons or offices often invokes questions of professional ethics. Especially in contexts of medical and research risks, consent is legitimated according to rules limiting the presumptive power of the authority consented to and restricting the kind of activities to which consent can even be given or received. Some approaches to exploring the dynamics of siting disputes rely on professional ethics models of consent to evaluate whether the harmful consequences of siting override or invalidate presumptions of informed consent on the grounds of an analogy to violations of professional ethics codes.[6]

This first perception of consent as involving deference to the authority of persons or offices is often an issue in risk decisions, especially when those decisions become public controversies. Although I refer to this notion of consent in this chapter, I primarily reserve discussion of the relationship between consent and authority in risk decisions for chapter three. It is often not easy to separate this issue of consent from a second, more subtle and foundational model of consent that is the primary focus of this chapter. In liberal societies, this second notion of consent springs from the understanding that some issues affecting an individual's welfare, liberty, or behavior are legitimately resolvable only on the collective level, rather than by autonomous individuals deciding for themselves. For instance, the solution to air pollution can be found only if society as a whole makes a collective decision and effort to clean up hazardous emissions. In other words, liberal societies are based upon an understanding of consent that says that on some issues individuals must—because they have consented to the existence of society itself—defer to collective decisions that invariably restrict individual choices.

It is my contention that the controversial aspect of risk decisions is primarily a result of the unwillingness of liberal citizens to acknowledge the applicability of this second sense of consent. The personal and proprietary NIMBY response to risk represents a refusal by an individual or group to consent to the redefinition of a risk issue as a matter of collective choice. That is, NIMBY essentially means I get to decide, not society as whole. This denial of consent threatens liberal society in its ability to cope with risks because it cuts deeply against the grain of liberal legitimacy as grounded in social contract. Such a denial also affects the relationship to established authority, of course, and so is not completely separable from the first model of consent to be discussed in the next chapter. But it is more threatening than the first precisely because it does not envision merely the election (or selection) of new authority figures; rather, it denies the legitimacy of collective authority as a whole. This chapter explores the relationship between technological risk and this second notion of consent as individual deference to the collective.

As we have seen in chapter 1, the emergent character of modern technological risks makes collective response to risks the only possible avenue for coping with them. Yet liberal citizens resist consenting to collective solutions precisely because they perceive that doing so threatens the loss of private goods guaranteed by liberalism's moral focus on the individual. Liberalism guarantees to individuals the moral right and autonomy to define for themselves what their good is to mean and furthermore bases this guarantee on the epistemological position that in fact individuals are capable by themselves of knowing where their good lies. Thus, achieving consent to collective definitions of the good is likely to be difficult, given liberalism's picture of the morally free individual.

This difficulty is further compounded by the introduction of tacit consent as a satisfactory mode of consent, to either persons as authorities or to the collectivity as a whole. Most models of risk assessment rely exclusively on tacit consent, and, as we will see in returning to liberalism's roots in our discussion of John Locke's version of consent, liberal societies in practice ultimately rely almost exclusively on tacit consent as well. Thus, policy making for risk manifests the core beliefs of liberal society as they relate to consent, as well as displaying the perilous state of liberalism as a legitimate model of social and political organization.

As opposed to express consent, tacit consent presumes some special aspects of the relationship between citizens, their society, and their government and complicates most issues of consent as they play a role in risk policy decisions. As a basis of political obligation, tacit consent has been a part of political theory since Plato chronicled Socrates's decision not to escape his fate in *The Crito*. Tacit consent became essential to the legitimacy and efficacy of liberal society with Locke's *Second*

Treatise on Government in the seventeenth century, and it remains a centerpiece of contemporary contract theories of political obligation and social justice, such as those presented by John Rawls and Robert Nozick. Construed as a legitimate foundation for political power, tacit consent has always been controversial, both within political theory and in politics, as current siting disputes show. A second goal of this chapter, then, is to explicate the tacit consent position to reveal both how it undergirds liberal politics and why it is under attack today.

One of the strongest arguments in defense of tacit consent, as we will see, is that it provides a moral argument for obligation, if not a completely persuasive political argument. It does so by stressing the relevance of the community within which political power is exercised against the individual. Tacit consent, I argue, is intrinsically a communitarian doctrine as communitarianism is construed in contemporary political theory (although Locke himself did not necessarily view it as such); thus, the threat to tacit consent today within risk politics invokes the whole debate between individualists and communitarians. It also makes clear why tacit consent is an "emergent" brand of consent, troublingly so for many individualistically minded theorists and citizens. Therefore, risk decisions lead us into a third discussion in this chapter, centering on the communitarian/individualist debate within contemporary political thought.

Consenting To Risk

Most approaches to risk incorporate the idea of consent into their understanding of what "acceptable risk" means. There are several reasons for doing so, beginning with the liberal background of risk decisions in the United States and other Western democracies and the role consent plays in that cultural environment. Liberal theory views consent as a normative requirement that, once met, can legitimate procedures or outcomes that have differential effects on populations. Of course, some models of risk assessment do not rely on consent, such as those invoking formal models of decision theory within a cost/benefit analysis or those stressing reliance on expert opinion. However, even in these approaches, there exists an assumption of a previous decision to use formal models or to rely on experts that itself depends on consent, usually derived through liberal democratic processes.[7]

More specific reasons also confirm the status of consent as a central feature of the acceptability of a risk. First, we are accustomed to being asked for our consent in personal risk situations, for instance, when facing surgery or before participating in potentially dangerous activities or sports. As I already pointed out, such requirements make up a substantial part of professional ethics codes, as well as legal doctrines incorporating strictures of informed consent. Second, as Douglas MacLean points out, many cases of risk taking already exist that formally require consent before socially beneficial action can be taken. Familiar examples here range from scientific experimentation involving human subjects to, since the end of the draft, the filling of military ranks exclusively with volunteers. Before such experiments or military training can proceed, detailed procedures must be followed that aim at ensuring that consent has been given, including tangible evidence such as signatures on forms. From a philosophical point of view, these are relatively unproblematic cases of consensual risk taking. Therefore, MacLean suggests, "we should take these simpler cases as our paradigm and proceed from there to see how the nature or the role of consent changes as the contexts of the risk decisions vary and become more complicated."[8]

Third, as MacLean also argues, consent plays a central role in our understanding of what risks are acceptable simply because other concepts do not work as well. Issues of equity, rights, or efficiency might well be raised in risk decisions, but it is not difficult to imagine situations in which each of them might be satisfied yet "the problems of setting and justifying an acceptable level of risk remain."[9] Of course, arriving at an acceptable level of consent does not guarantee that these other issues are immediately resolved. Indeed, we will see in later chapters how many technological risk controversies turn specifically on issues of rights or justice when consent does not apply—for instance, when considering the impact of present risks on the lives of future generations.

Here, I first examine the three reigning models of risk assessment, all of which construe the issue of consent somewhat differently yet highlight its centrality in risk decisions. They also jointly confirm that the type of consent at play in these decisions is always "tacit" in some sense and to some degree. Finally, they also agree that the recipient of consent is society as a collective body, since none of these approaches identifies persons (or "authorities") to whom consent is being granted. Even within the so-called "expressed preference" model, the tacit nature of the consent is obvious once it is generalized to cover all of society. The three approaches are two examples of "implied" consent, in MacLean's terminology, first, the revealed preference approach and second, the expressed preference model; and third, an example of what MacLean calls hypothetical consent, the "natural standards" model.

Revealed Preference as Consent

The revealed preference approach to risk assessment begins with the assumption that any society's level of acceptable risk is not something that can be either derived or observed in a short interval of time. Rather, societies gradually arrive at an acceptable balance of risks to benefits through a process of trial and error during which some risks are accepted by societies and some are rejected. Thus, in answer to the question, "How safe is safe enough," Chauncey Starr modeled an approach embodying two basic assumptions. First, measurable data indicating past acceptance of risks—Starr used data from historical national accident records—"are adequate for revealing consistent patterns of fatalities in the public use of technology." Second, "such historically revealed social preferences and costs are sufficiently enduring to permit their use for predictive purposes."[10]

In terms of consent, clearly Starr's approach makes several assumptions. First, it assumes that consent has been given to certain risky activities in the past. If they were not in fact consented to, there would be no reason to assume that they now form a consensual basis of acceptable risk. This consent may or may not have been expressed, although if it is construed as expressed, one must assume that certain activities (e.g., buying a car and driving it) constitute consensual expressions. Second, if such instances of (possibly expressed) consent are to indicate consent to other risks in the present, then whether past consent was expressed or not, current consent is obviously tacit (e.g., as manifested by my continued driving of a car). Furthermore, this brand of tacit consent is especially metaphoric in that the comparable activities in the past and present may be very different in every respect other than in amount of risk. For instance, an activity in the past that currently carries the same amount of risk as driving a car may not resemble in any other way the activity of driving. This of course raises problems of isomorphism in the comparison—how can such different activities be comparable simply because one of their elements, risk, happens to be?

Starr's approach has been criticized on several grounds, some of which appeared in the first chapter. Baruch Fischhoff and his colleagues, and others, point out the excessive conservatism of the revealed preference model. This conservatism takes several forms. First, the model assumes that society has already reached an optimal balance between risks and benefits.[11] The presumption here against change is odd, to say the least, since we are speaking of technological risk and development, which would, one assumes, mean change and a great deal of it. Second, in assuming then that the past level of risk acceptance establishes a normative limit on current risk, Starr subscribes, Kristin Shrader-Frechette asserts, "to a highly doubtful presupposition, namely, that values don't change, and that societal norms are not dynamic."[12] Third, as Robert Goodin pointed out, past levels of risk acceptance carry distinctive patterns of distribution both of risks and benefits, and these may no longer be acceptable to a society on other than economic grounds. For instance, if a disproportionate share of risks were taken in the past by the poor, by members of certain races, or by women, society today might have moral grounds on which to reject the past standard of risk acceptance.[13] In the United States at least, the histories of the civil rights and women's movements have proceeded precisely on such normative claims about the debilitating nature of racism and sexism. Thus, for example, the risks associated with slavery, ghetto life, or medical science's past treatment of "women's problems" are ethically unacceptable today. Successes in both movements can be partly measured in terms of society's rejection on normative grounds of certain risks that were limited in their effects to either of these groups of citizens.[14]

The implicit conservatism of the revealed preference model therefore undermines its usefulness as a predictive device. This is especially true when it comes to assessing the acceptability of technological risks. Modern technological societies almost by definition are changing at a great rate in part due to the technologies themselves. As these technologies grow and spawn new areas of research and development, does it make sense to use the risk assessments of earlier, nontechnological periods?

In terms, then, of the consent that citizens are presumed already to have given, can we really presume to tell citizens that if they accepted the pollution of coal-generated energy, we can assume they also accept the cleaner, but potentially more catastrophic effects of nuclear-generated energy? Are these two even comparable as risks? Could we tell someone accustomed only to travel by horseback (and therefore accepting of the risks of being thrown) that the risks of jet travel are even fewer and therefore he or she has by past acceptance already consented to them? The nature of consent here is not only exceedingly tacit, but also too irrevocable—at what point can one object to additional risk that is less than that accepted in the past and on what grounds? Such questions reveal the weakness of the revealed preference approach as a model for resolving political issues surrounding risk, because it grants little leeway for either the rescinding of past consent or for denying its applicability to allegedly analogous cases in the present.

Expressed Preference as Consent

As a second "implied" consent model of risk acceptability, the expressed preference approach shares some of the same weaknesses of the revealed preference school of thought. There is a significantly different starting point with the expressed preference approach, however. Rather than, as in the revealed preference model, simply inferring from society's past performance its consent to risks, this model sets out to ask people directly. Relying most often on opinion surveys, this approach seeks to assess the relative importance of certain attendant aspects of particular risks, such as voluntariness, uncertainty, chronic versus catastrophic nature, equity of distribution, and the like. Second, it analyzes the results in terms of the rationality of the preferences expressed. Third, it makes recommendations concerning both the acceptability of risks and the education of the public in order to make more effective risk policy.

As carried out by some of its most prominent practitioners, Baruch Fischhoff and his co-investigators, this approach arrives at impressive findings concerning the public mood about risks in the past and present and its willingness to consent to additional ones in the future. The expressed preference model does, however, fall victim as a policy-advising device to two failings. First, it suffers from the usual difficulties arising from the use of surveys. These come from the shortcomings of its intended target group, who are, given the vicissitudes of random sampling, all of us. As Fischhoff's various studies make clear, the public frequently is too ill-informed, uninvolved, or even cynical to make entirely rational decisions concerning the acceptability of future risks. As a result, its preferences may be the result of ignorance and may exhibit features of inconsistency and even incoherence in its reactions to certain risks. Besides, surveys have a short shelf life. During campaigns, for instance, pollsters are careful to explain that any particular survey is not a prediction of a distant election day outcome but only a "snapshot" of the current leanings of the public.[15] Risk politics shares the volatility of electoral politics; thus, any expression of preference concerning the acceptability of risk as evidenced in surveys should not be presumed to have long-term validity. Therefore, assuming the public's consent on the basis of such responses is, at the very least, a dangerous political gambit.

A second problem with the expressed preference model as a prelude to policy is (as Fischhoff and his colleagues are aware) the sneakiness of its version of consent. Because people are actually being asked questions about their attitudes toward risk, it is easy to assume that actual expressed consent is being sought, and possibly given. This, however, clearly is not the case. As in any survey, only a small sample of the citizenry is being asked about its likelihood of consent, not the entire electorate. Thus, even if it is correct to assume that predictions can be made about public response on the basis of these surveys, this surely cannot constitute consent in any more than a tacit form. In fact, this model depends on even more "tacitness" in the definition of consent than does the revealed preference model, for it does not presume to observe even the whole of society, but only an (at best) random sample.[16]

Furthermore, the findings from research such as that of Fischhoff and his associates makes it even more clear that tacit consent is an insufficient guide in risk decisions affecting society. If, as they argue, their sampling of citizens demonstrates considerable misinformation, cynicism, or even irrationality, we should certainly not take any ascription of consent arising out of that survey as indicative of more widespread consent. Consent based on such levels of negative inputs could hardly meet legal or ethical standards of "informed consent" in the "unproblematic" cases MacLean suggested as paradigms for our more difficult decisions. Therefore, it could hardly be a basis for widespread social policy concerning risk.

Natural Standards as Consent

Both implied consent models of risk assessment require their practitioners to make judgments about the values at work in society as a whole and to assess their relevance to the public's attitudes toward risk. Since doing so is always difficult—and sometimes leads to fallacious conclusions—there is an intuitive appeal to ignoring people altogether and taking nature as our guide in determining which risks are acceptable. The natural standards approach does this by offering geology, biology, or other of the "hard" sciences as an alternative to the psychological and political uncertainties of the implied preference approaches. As Shrader-Frechette says, "these geological and biological criteria specify the levels of risk which were current during the evolution of the species." Relying on them as standards, then, would presume that "the optimal level of exposure to various risks is that which has naturally occurred."[17]

The appeal of this approach lies in its apparent ability to ignore differences in culture across different societies and therefore to offer generalizable and quantifiable conclusions about which risks are beneficial and which are detrimental. Two examples where use of this model seems efficacious are in the setting of acceptable levels of radiation or of lead in food. Both standards would be equivalent to those levels found in remains of humans uncovered during archeological research.[18] The assumption here then is that if the species as a whole had thrived within certain limits of risk exposure, then consent could be presumed as a matter of adaptation and survival. The US Code of Federal Regulations sets the latter limit of "background radiation" as not to exceed 500 millirems.[19]

Despite its scientistic appeal, the natural standards approach has several failings as a model of risk assessment. First, if it is used (as it often is) to argue for accepting any risk of less than natural standards, it is a dangerously fallacious argument. Robert Goodin comments that advocates of nuclear power often point "with glee" to the fact that risks associated with that technology are considerably less than those of, say, crossing a busy street. They miss the crucial point, however, that such risks are cumulative; therefore, willingness to accept one of them is no warrant for assuming that all other risks of equal or less magnitude should logically be accepted. Goodin writes:

> The standard practice of using "natural background" levels of radiation as a benchmark for judging how much more radiation may be introduced into the environment is utter nonsense. Even if people have reconciled themselves to suffering this inevitable background radiation, that is no reason to suppose that they would be willing to take double the dose, which is what we are really contemplating when asking them to subject themselves to as much again from a manmade source.[20]

Second, the apparently scientific basis of the natural standards approach does not bear close scrutiny. The prejudices of allegedly "natural" standards have often been pointed out by some philosophers of science who see in them either male or race bias. It is conceivable that some natural risk standards that appear to warrant universal applicability may be different for women than for men (as in the case of certain gender-specific forms of cancer) or for people of different races or ethnic groups (as in the case of population-specific medical maladies such as Tay Sachs or sickle cell anemia). We might, it is true, rely on science eventually to ferret out these nonscientific "natural" standards, as it has in these cases. Nevertheless, the history of science is replete with examples of science's failure to do so before discriminatory practices have been wrapped for decades in the cloak of scientific legitimacy.

Furthermore, this approach shares in the excessive conservatism of the first two models. It is not difficult to generate examples of new substances for which there are no historical natural standards. Saccharin or aspartame are good examples here, because neither exists anywhere in nature. More menacing perhaps are the yet to be "discovered" products of recombinant genetic technology. The wonders likely to be produced by this new technology are only now beginning to make their way into the marketplace, the hospital, and the public's mind. Scientists and governmental agencies are treading cautiously in the development and approval of genetically engineered substances, it is true, but the natural standards model has nothing to offer by way of guidance, for none of these products has ancestors, or in some cases even analogs, existing in nature. The model's silence on these substances (that are very much with us) limits its usefulness in an age when new life forms are not only created with disconcerting regularity but patented as well.

Other criticisms of the natural standards approach are possible and have been made by others.[21] For our purposes here, however, one argument against it is most telling. In its ability to ignore "the human factor" in risk assessment, it also ignores the idea of consent altogether, as well as some important features of human evolution. Or rather, the natural standards model assumes that we "tacitly" acquiesce to the environmental background conditions within which we find ourselves. Focusing on such conditions neatly sidesteps any issue of what is being consented to, but goes excessively far in its removal of humans from the risk equation. From an evolutionary perspective (which this model adopts as its standard) the argument becomes comically ill-considered if one views human evolutionary development as a whole. If acquiescence to background conditions were really our species' way of granting consent, how did we ever progress beyond our cave-dwelling stage in the first place? The history of our species' development is mostly about the refusal to accept background conditions. How else does one explain the human development of science and technology if not as an evolutionary attempt to overcome certain limitations within the "background conditions" of human existence? It is surely *because* of our species' stubborn refusal to accept such background conditions that the urge to science (as well as to justice and liberty, among others) partially arises. Whether we proudly or sheepishly acknowledge the Enlightenment's motivation for science as a tool for conquering nature, it seems pointless to deny its efficacy in the history of Western culture. We look to science precisely to help us overcome our "natural" limitations, both in ourselves and in our environment. It is the reason we assume the risks of discovery. It is also a reason we insist on our continued (and expressed) consent to the progress and practices of scientific research and technological development.

Much more has been and could be written concerning each of these models of risk assessment. For our purposes, however, it is sufficient to draw two conclusions from them of relevance to our larger purposes in this chapter. First, the concept of tacit consent is a central feature of all these models of decision making for risk. This does not necessarily mean that actual expressed consent is never warranted by any of these models or that they would never welcome it. But it does mean that each of these models is an attempt to provide a way to make decisions about risk without *requiring* expressed consent.

Second, these models presume that no further consent is necessary since the consensual basis of current social practices surrounding risk are validated by past acceptance of the same or similar practices as a matter of consenting to society as a whole. That is, these models assume that whatever consent new risks require has already been provided by some initial consent to either liberal society as a whole or the authorities that it encompasses. Risks are the consequences of collective life; thus, to the extent that collective social and political life has been accepted, new risks, especially since they are the emergent products of our collective lives, require no further grants of consent from citizens. Thus, these models validate a decision methodology that allows policy making to proceed without delay for citizens to proffer their consent. Waiting for citizens to actually make up their minds and express themselves on risk issues is surely cumbersome and not itself without political risks. Primary among them is the possibility that requiring more than tacit consent will make consent more difficult to attain and policy therefore more difficult to generate.

These possibilities, of course, define the policy quagmire that the NIMBY syndrome has produced within the contexts of several different technologies. Relying as it does on a presumption of tacit consent as generated by any of the models of risk assessment we have explored, policy making for technological risks has been impeded on several fronts by citizens who deny that tacit consent is a sufficient warrant. Whatever else they entail, claims of NIMBY are at bottom demands that only expressed consent can legitimate collective action regarding risks. As "expressions," NIMBY behaviors are themselves public acts meant to demonstrate that tacit consent is not enough for those who say "you can't put it here."[22] In their failure to recognize the essential unpersuasiveness of tacit consent, the three models of risk assessment discussed here cannot prepare either policy makers or the political process as a whole for the powerful—if negative—impact of citizen activism.

This lack of foresight by models of risk assessment is not surprising, however, because liberal societies as a whole ground their definitions of political legitimacy primarily on models of tacit consent. This only goes to prove that historic reliance on tacit consent as a legitimator provides the reasons why liberal societies cannot cope with technological risk and it is why NIMBY claims are more than selfishness or egoism; they constitute a fundamental attack on liberalism in the message they carry "about who should rule and what should matter."[23] Finally, the damage done by the NIMBY syndrome is also proof of the unworkability of tacit consent as a basis for political obligation. Thus, the citizen response to technological risk decisions challenges the consensual foundations of liberal society as a whole.

Consenting to Liberal Society

Carole Pateman explains the liberal dilemma presented by the NIMBY response to risk policy in her assertion that political obligation in the liberal state constitutes an "insoluble problem" of consent. "If consent is to exist in 'some significant sense' then, at a minimum . . . consent has to be seen as something that arises from the acts of citizens, [not] as a matter of governmental manipulation."[24] This requirement seems on its face to rule out *tacit* consent as the source of political obligation, since tacit consent does not presume any particular "acts" on the part of citizens. But, furthermore, as Pateman discusses at length, as explored here in chapter six, liberal democracies do not incorporate enough *acts* of participation within their governing process that would count as actual *expressed* consent. (Unless, of course, one counts NIMBY episodes. But because these are by their nature a denial of consent and obligation, they cannot be counted as acts that confirm consent. They do, of course, confirm that the opportunity to consent is being demanded.) Therefore, in her view, the presumption that political obligation has a basis in consent in liberal states is unfounded.

But is tacit consent so weak a warrant for obligation? In one of its earliest formulations, tacit consent seems to carry a reasonableness and sense of fair play that recommends it as a basis for duty. Locke presses his famous case for obedience:

> There is a common distinction of an express and tacit consent which will concern our present case. . . . The difficulty is, what ought to be looked upon as a tacit consent, and how far it binds—i.e., how far any one shall be looked upon to have consented and thereby submitted to any government, where he has made no expressions of it at all. And to this I say that every man that has any possessions or enjoyment of any part of the dominions of any government does thereby give his tacit consent and is as far forth obliged to obedience to the laws of that government, during such enjoyment, as anyone under it; whether this his possession be of land to him and his heirs for ever,or a lodging only for a week, or whether it be barely traveling freely on the highway; and, in effect, it reaches as far as the very being of anyone within the territories of that government.[25]

Locke can safely rely on tacit consent partly because of his prior insistence on natural laws that operate even before the existence of government and that limit the amount of risk that individuals feel in his version of the state of nature. These "natural standards" resonate within our moral intuitions even today. Indeed, the echoes of Locke's argument are very much a part of the natural standards model of risk assessment, as they are of contemporary liberalism as a whole.[26]

Locke's argument relies on tacit consent even before the whole issue of political obligation arises, that is, in the state of nature before the creation of government. Tacit consent is the basis of the invention and acceptance of money as a means of exchange. Because property (and the natural right to it) already exists in the state of nature, Locke initially argues that the laws of nature limit property's acquisition to how much any single person "tills, plants, improves, cultivates, and can use the product of."[27] Obedience to this injunction results in a rough equality in distribution. Nature limits the amount of property any one person can own in two ways. First, the nature of property itself militates against excess acquisition and therefore inequality—things that become property are either subject to spoilage if they are unused, or, in the case of land, subject to wasted effort. So nature teaches the irrationality of surpluses. If surpluses are irrational, the various differences between people in their abilities to amass surplus goods (such as superior strength, stamina, or intelligence) are irrelevant in how nature's bounty is distributed, so they go unexercised.

Second, the "Lockean Proviso" expresses the normative side of the law of nature that in acquisition one must leave "as much and as good for others." The invention of money of course nullifies the first, naturalistic limitation on acquisition by making surpluses useful. What is somewhat unexpected is that Locke accepts money and the "disproportionate and unequal possession of the earth" it immediately creates as morally legitimate since they are the result of "men tacitly agreeing" to them.[28]

The introduction and acceptance of money immediately breeds inequality based upon differences between individuals that are now relevant to the distribution of property. This consequence appears to violate the natural laws that Locke has viewed as operative in the state of nature and that guarantee to individuals, among other things, their "lives, liberties, and estates, which I call by the general name 'property'."[29] This apparent disjunction has been much commented on by Locke scholars, who disagree on whether Locke's acceptance of the right of property and the use of money ultimately violates the natural law.[30] If it does (and probably even if it does not), then the single *moral* foundation for social interaction (the proviso) is being exchanged for another (the legitimacy of inequality). What results then is a new moral order initiated not by the natural law but by tacit consent. Tacit consent gives rise to a new moral community.

The prior consenting to money—and its widespread social, political, and normative ramifications for Locke's theory—illuminates two important aspects of consent theory as a whole and raises as well questions of how to interpret Locke's use of tacit consent. First, regardless of whether the consent to the use of money is construed as tacit or expressed, it is clear that such consent presumes an already existing moral community within which individuals agree on certain rules and practices;

in Locke's state of nature, those rules and practices are demarcated by the laws of nature. Significantly, however, it is clear from Locke that at the point at which money—and the resultant moral order legitimating inequality—is consented to, government does not yet exist. Thus, the obligations entailed by participating in the moral community constructed by tacit consent are not, strictly speaking, political obligations, because political relations, at least as Locke subsequently perceives them, have not yet developed. This is important for our understanding of the nature of liberal community and liberal obligation. Tacit consent both presumes, as Craig L. Carr writes, "certain normative relations holding amongst persons that were not the result of an explicit contract" and is the basis on which "the manufacture [of] social conventions that embellish the otherwise sparse normative relations that set the basic duties persons owe to each other under natural law."[31]

Thus, for Locke tacit consent presumes the existence of both obligation and community and also provides the means by which both can be transformed from moral to political forms. I will illustrate shortly the logical difficulty of that transition from moral obligation and community to political obligation and the state for both Locke and consent theory generally and explain why it is of crucial importance in assessing the obligating potential of tacit consent in risk issues.

A second important aspect of Locke's construal of consent in the state of nature as legitimating the new moral order ushered in by money is that it lays the groundwork for a later political obligation based, in Locke's view, upon expressed, not tacit, consent. Because Locke makes it clear that the primary purpose of government is to adjudicate disputes concerning property, we must assume that since those disputes did not exist before money, they are the direct product of its invention. Thus, government is a necessity but requires the social contract—an opportunity for expressed consent. This is a new type of obligation in the social world, one that legitimates the exercise of power in coercive fashion by representatives of the government.

Even granting Carr's point that Locke's understanding of political obligation as based upon tacit consent is conventional by seventeenth-century standards, Locke realizes that this new form of obligation—political—requires more to legitimate it than did the moral obligation governing normative relations in the state of nature. In short, this new obligation must be based on expressed consent. So he makes two arguments. First, Locke claims that those signing the contract are clearly expressing their consent. Second, as David-Hillel Ruben makes clear, acceptance of the protection of property made possible by government now constitutes for Locke not tacit consent, but actual expressed consent:

> But how can the descendants of the original contractors promise explicitly if they do not sign the contract as did their forefathers? The answer is that, for Locke, owning property via inheritance is not tacit promising, which could in any case only amount to the obligation of a resident and which could make one merely a resident of the state. Owning property via inheritance is an alternative way of explicitly promising, for only as such could owning property confer membership in the state.[32]

But this acceptance of property protection will not wash as expressed consent, at least not to the extent necessary to legitimate Locke's liberal society as a whole.

Why not? It simply denies to too many citizens the opportunity to express their consent. In Locke's time of course the exclusion would extend to women, slaves, debtors, and, in most cases, offspring not fortunate enough to be firstborn. Locke might well accept the exclusion of women and slaves from the ranks of full citizenship (defined as those with the opportunity to express consent); at least this is debatable. But if Ruben is correct in his interpretation, Locke would have to exclude even all male offspring not eligible to inherit—that is, those not benefiting from primogeniture. Alternatively, if Locke opposes primogeniture and leaves inheritances to the will of the legator, then citizenship for children would be at the discretion of their fathers. Locke clearly opposes this, as his first treatise makes clear, as does any notion of citizenship by consent. Even if Ruben is correct, then, that Locke means to broaden the definition of expressed consent to include those who inherit, this does not in itself transform consent from tacit to expressed for all—or even a majority—of citizens.

Although Locke is one of the first contract theorists who seems to require expressed consent at least to the contract, the fact remains that the main function of government is to protect property rights and therefore also the unequal property distribution that tacit consent legitimates before government is even on the scene. Thus, it seems that the obligations arising from tacit consent both predate and subsequently determine the nature of the obligation arising from the later expressed consent to the contract, such as it is.[33]

Ultimately, then, Locke's understanding of moral and political obligation is based, with a small number of property-holding individuals as exceptions, almost entirely on the notion of tacit consent. Locke's reliance on tacit consent is paralleled by that of other social contract theorists, both historical and contemporary. Rousseau's account of the state of nature traces a long series of events involving the growth of inequality that are all either "random events"[34] or the tragic results of a tacit acceptance of growing social dependence. Similarly, John Rawls and Robert Nozick engage the reader in "thought experiments," transporting him or her back to an original position or natural state within which decisions are made about issues such as justice or liberty that define the present moral order. In all these theories, the purely hypothetical contract never completely satisfies as an argument, since, as Will Kymlicka says, it relies "on very implausible assumptions,"[35] or, as Ronald Dworkin remarks, echoing the eighteenth-century position of David Hume, such a tacit contract "is not simply a pale form of an actual contract; it is no contract at all."[36]

Still, though social contract theory does a poor job convincing anyone that expressed consent was ever widely present or that tacit consent is enough to fully legitimate political power, it often carries considerable power as an (admittedly) intuitionist argument for moral obligation. All consent theorists, as Carr makes clear,[37] implicitly or explicitly take the contract as constitutive of a community based on normative principles of duty and obligation. These foundationalist principles may differ—from negative liberty and natural rights in Hobbes, Locke, and Nozick to participatory engagement in Rousseau to a vision of social justice in Rawls—but they all assume that a moral community results from their acceptance. Pateman correctly critiques all such theories, however, noting that even though contract theories might succeed in instilling a "horizontal" obligation between citizens, they cannot turn this into a legitimate "vertical" obligation to obey government.[38] In other words, liberal (i.e., tacit) consent can create moral obligation but not political obligation.

This difference between moral and political obligation, and the fact that tacit consent can only legitimate moral obligations, is a crucial point in understanding why NIMBY claims have power and have successfully stymied policy related to risk. All such claims presume a moral community that can recognize the claim as a statement of moral obligation, yet they reject the political obligation to acquiesce to what is being proposed as a matter of public good. In other words, liberal consent theory can succeed in creating an effective moral community but has trouble adding the element of political power into the mix in what would be perceived as a legitimate manner. Thus, the political legitimacy and power necessary to make policy is simply lacking. As an example, consider the following version of the old state of nature metaphor, updated to our time as a hypothetical situation, as is the practice of contemporary state of nature theorists.

Sunday Afternoon in the State of Nature

On a warm spring afternoon in a New England college town, an undergraduate leaves her dormitory for a walk. It being Sunday and New England being itself, her rambles take her through a deserted town green with closed shops and quiet streets. Walking into the woods on a damp path and across a narrow footbridge brings her past stone-walled cemeteries to a small town park she had never visited before. Much to her surprise, the park is filled with music and bodies swaying in rhythm in front of a band performing on a raised stage. She joins a group of about 20 college-age people listening and dancing to the music. She is astonished by the event, unadvertised on campus and, by Sunday standards in New England, relatively energetic and, well, fun.

During the course of the afternoon the student makes some new friends among those who already seem to know each other well and thoroughly enjoys herself. Late in the day as she sits on the grass, one of her new acqaintances comes by with a plate piled with sandwiches and a pitcher of beer. She is offered both, and, since her dorm cafeteria does not serve supper on Sundays (and never serves beer), she accepts, congratulating herself on her good fortune. She leaves the park late in the day aglow with luck and new friendships and makes a mental plan to return every week. For the next month she is a regular at the concerts, enjoying the music and the food. Each week different members of the group perform on stage—always to great effect—and other members appear to take turns distributing delicious refreshments.

On one subsequent Sunday in her secret musical park (as she thinks of it), supper comes with a tab. "We're so glad you've joined us," a young woman in a peasant dress exclaims. "As you have no doubt noticed, our practice is to share food with everyone present, as long as they recognize their duty to return the favor; next week is your turn to bring the food." Although she silences her immediate and probably rude noise of response, several thoughts rush through the undergraduate's mind. "I don't remember 'joining' anything," she mentally proclaims, "and I certainly never bargained for having to feed all these people myself. I'm not bringing any food next week, and no one can make me." Dimly recalling a metaphor from some distant political theory class, she forces a smile and walks home.

It is not difficult to anticipate Locke's answer to the question of whether the freeloading student is now obligated to bring the food for next week's concert. Because she has entered an established community containing both rules and benefits, that, with even the smallest reflection, would have been evident to her, and the latter of which she has enjoyed, she is understood by Locke to have given her tacit consent to the rules according to which her benefits have accrued. She is obligated to reciprocate, do her part, obey the rules, however her obligation is construed. Not to do so makes her . . . what? This is the real question, of course, summed up (but not really answered) by Locke when he asks how far tacit consent binds. If she does not cater the next concert, is she a criminal? A moral reprobate? Merely a bad empiricist (since she should have better noted the mechanics of the community's operation)? A person of great promise in the modern business world? Although Locke clearly sees the moral obligations derived from tacit consent as more than those merely of fairness or gratitude, it is not clear how far they extend.[39] For Locke that clarity is what is lacking in the state of nature, and the reason why it must be left behind.[40]

As difficult as these questions appear, there are others as well. For instance, what if our undergraduate does not supply food as she is commanded; can she nevertheless return to the park in the future, or are her hungry friends entitled to deny her the pleasure of their music? It is a public park, remember, belonging to the town and open to all citizens, regardless of their musical or culinary predilections. If her friends do not deny her entry, must they then feed her as well? After all, she has grown accustomed to their largesse and could even conceivably claim that as a penurious student she is entitled to their assistance on days when no other food is available. Does she have a claim to both entertainment and a meal?

Besides such free rider and entitlement issues are additional questions of a different sort. If our student proclaims her intention to avoid bringing food, are her newfound friends (and benefactors) entitled to compel her physically to meet what they consider her obligation? Can they punish her if she successfully resists? What if the absence of compulsion or punishment results in more of their members emulating our student's deadbeat ways to the extent that the community itself fails? Would this eventuality (or its clear and present threat) legitimate the collective exercise of coercion against our student?

These kinds of questions accompany any state of nature or social contract argument, and, indeed, they constitute a large part of introductory college ethics texts. It is important to recognize, nevertheless, that these questions are of two distinct types. It is a failing of Locke's theory that he conflates the two in his reliance on natural law and reason. Locke's mistake is rarely made by citizens when pressing their NIMBY claims.

The questions raised by our undergraduate's self-centered behavior invoke two types of obligations it is essential for human liberty to keep separate: those required for the health of a moral community and those needed for perpetuation of the state. Moral obligations and the rules pertaining to them create and keep communities together.[41] It is the agreement on these throughout a community's life that makes it recognizable as a community. These are often the product of oral history and shared tradition rather than expressed consent, and violation of them can lead to disapprobation, even censure.

Political obligations, on the other hand, are not reducible to these moral obligations, although, according to some scholars, Hobbes tries mightily to effect this reduction in his argument for why the social contract cannot be avoided.[42] Political obligations add a new element—institutionalized power, thereby codifying some obligations into enforceable law but, in liberal society, leaving others to the discretion and moral intuitions of individual citizens. In its foundation in "higher law," liberal society recognizes the distinction between the duties owed to the state and to others, whether the others are identified as humanity, God, or even oneself. It is where the right of dissent arises.

Since Locke is one of the first to recognize the right to dissent—and even to revolt—it is curious that he fails to distinguish between moral and political obligation. But he clearly fails to do so in his insistence that acceptance of benefits implies consent and therefore also conveys obligation to obey the law. In our example, we might be persuaded that the student has a moral obligation to bring food to the next concert, even that she is a reprobate (or a "sleaze" in today's parlance) if she does not. But our moral disapprobation does not automatically grant to the members of her newfound musical community the legitimate power to physically force her to feed them. Legitimate coercion arises from an obligation of a different, political sort—and one that it is simply too risky to presume she has voluntarily accepted simply by the acts of remaining in the park or of accepting food. Tacit consent, in other words, cannot be taken as the basis of a legal contract or of political legitimacy. Confusing moral with political obligation is a denial of choice between alternatives, which are often not nearly so clear-cut as in our example. Therein lies the liberal fear of a too powerful state or a too embracing public realm. It is within this dilemma of obligation that Pateman's argument that liberalism cannot account for political obligation really begins, and where it succeeds.

We should be careful, nevertheless, not to belittle the power or importance of moral obligations nor the degree to which they bind people together. Indeed, the success of NIMBY claims is due to the shared moral sense within liberal society that the claimants have good moral grounds for denying the violation of their private space by the state's plans to build something in their backyard. Those claims, being liberal, are most often based on conceptions of the individual's rights at issue, since "rights" are the traditional moral currency of liberal society. Use of that currency empowers individuals to say "no" to the collective; rights provide the grounds for accepting obligations as moral but rejecting them as political. As citizens of the liberal state, we recognize the right of a fellow citizen to say "not here," and we deny on grounds of *moral* obligation the right of the state to overcome that objection on *political* grounds.

Our example also shows the weakness of the concept of political obligation to the liberal state; therefore, that state's inability to make effective policy for technological risk in the face of NIMBY claims is not very surprising. Given the ingrained preeminence of moral obligations within liberal society and liberal theory—even in

Locke—political obligations in the sense of what one owes politically to others simply are a hard sell to most citizens faced with the kind of negative risks that modern technology poses. When confronted with a political decision, for example, that will site a radioactive waste dump next door, many citizens have rejected any obligation arising from the political argument that they should accept the risks involved as a matter of good citizenship or of serving the public good. Such arguments simply do not carry sufficient weight in citizens' minds, at least in comparison to the arguments invoking moral obligations (for instance, to oneself or one's children placed at risk), and, because they do not, siting disputes erupt and policy making founders.

The weakness of political obligation in the liberal state goes directly to the insufficiency of tacit consent as a warrant for subjecting citizens to the risks of high technology. If political obligation is construed as a horizontal obligation owed to other citizens (as in Walzer's formulation), citizens have little difficulty in proving other horizontal obligations of a strictly moral nature that supersede the political claim. Usually these are obligations owed to those closer to the claimants, often family or near neighbors. Thus, as David Hume pointed out in the eighteenth century, moral obligation binds most when distance between those feeling the obligation is least.[43]

Compared to the political obligation to let one's hometown be one of two or three sites for dumping radwaste for the entire nation, the welfare of those closer to home naturally seems to exercise a stronger obligatory tie. Why does it? Because in our everyday activities we are involved in our family's and neighbors' lives in a way and to an extent that is not true for our interaction with fellow citizens thousands of miles away. As we will see in chapter five, our actions have impacts on them that are stronger than on those further removed from us. It is the impacts of our actions on them (and theirs on us) that bestows greater moral force to the obligations owed our proximate neighbors. This does not mean only that we "feel" a stronger obligation. It *is* stronger because it is visibly based upon activities and actions that can plausibly be taken as evidence of *expressed* consent to a relationship of mutuality. In other words, in a tightly knit community, actions that imply consent are both more visible and visibly consensual. Therefore, they present stronger arguments for a presumption of obligation. Our student in the park, for instance, seems especially reprehensible in her refusal to recognize her obligation because she has a history of close interaction with her small group of new friends. They who require her food are not the faceless, homeless crowds in distant cities or of future generations—they are people with whom she shares a current and close relationship.

The matter of recognizing sufficient political obligation to cope with technological risks becomes even more difficult if the obligation is seen as a vertical relationship between citizen and government. As Pateman argues, the avenues for political participation in liberal states are simply too few to mirror the degree of involvement in other people's lives that might give rise to a presumption of moral obligation, as in small communities. It is often the activity that matters—the doing and the doing together that builds the sense (and moral persuasiveness) of obligation.[44] Because modern liberal democracies presume little political activity on the part of citizens—and what there is most often takes place in the privacy of a ballot booth—they do a poor job of constructing a plausible sense of vertical obligation to obey those elected to make policy decisions.

The issues of participation and authority within risk policy decisions will arise again in later chapters, but both are part of the general problem of obligation within liberal society. Liberalism since Locke has recognized moral obligations based on a shared (but nevertheless tacit) consent as more powerful than political obligations that allow coercion in the name of the public good; this is the moral basis for the liberal principle of limited government. Thus, it should not be surprising that citizens today manifest this same set of moral and political principles. One additional reason, however, goes to the nature of community.

Liberalism's reliance on tacit consent proves that it is not completely inured to the necessity and meaningfulness of community. It is through tacit consent that the moral community of liberalism is manifested and evolves, even if, as in the case of Locke, that evolution proceeds upon tacitly consenting to the inequality wrought by money's displacement of the Laws of Nature. But it is a view of community that falters as a political vision because of the demand that power must be exercised on the basis of a stronger view of consent—expressed consent. Without the vision of a truly political community, then, liberal societies are impotent in the face of the challenges of technological risk. Making policy to meet these challenges requires two things from citizens: first, a recognition of their strong connectedness to their fellow citizens in the face of the emergent nature of risk and, second, an acceptance of power's necessity and hence of political obligation to respond to that risk. These are two communal obligations that liberalism has traditionally had difficulty in laying at citizens' feet, much less in placing around their necks.

Consenting to Community

Liberalism has been accused for 200 years of, at best, ambivalence and, at worst, antagonism to the ideal of community.[45] The accusation can be traced to Locke's view of the social contract and its ambiguity concerning moral and political obligations to others. As Michael Sandel points out, contract theories generally embrace two distinct values: individual autonomy and reciprocity. As I have argued, the two ideas of consent—expressed and tacit—correlate serially with each (but only one) of these values, and the obligations arising from tacit consent are dominant, at least in Locke's social contract theory. Sandel says:

> [W]e may think of the morality of contract as consisting of two related yet distinguishable ideals. One is the ideal of autonomy, which sees a contract as an act of will, whose morality consists in the voluntary character of the transaction. The other is the ideal of reciprocity, which sees a contract as an instrument of mutual benefit, whose morality, depends on the underlying fairness of the exchange. . . . Where autonomy points to the contract itself as the source of obligation, reciprocity points *through* the contract to an antecedent moral requirement to abide by fair arrangements, and thus implies an independent moral principle by which the fairness of an exchange may be assessed.[46]

In my reconstruction of Locke, reciprocity within the society of the social contract is based on a moral idea—that consent is necessary to legitimate inequality;

thus, it is in fact a moral community founded in tacit consent. Yet Taylor and other communitarian critics are also correct that the reciprocity of the liberal society is not very communal because it is based mostly on shared economic interests—traceable to the initial cause of the reciprocity, the tacit agreement to employ money. Communal commitment can extend only so far because of the other value present in Locke's liberal theory—autonomy. Autonomy demands that further extensions of communal obligations be founded in expressed consent; thus, for Locke the social contract as embodiment of that consent was logically required. To the extent, then, that expressed consent can never really be present in liberal society, the extension of communal commitment that Taylor seeks cannot occur.

Liberalism caused a revolution in political thinking (and several in practice) because of the concept of individual autonomy. From it come the ideas of inalienable rights (especially the right of privacy), the division of life into public and private realms, the idea of limited government, and the overarching goal of personal liberty. In the behavior of citizens, autonomy dictates two things: the recognition of certain basic moral obligations to others and the denial of the transference of autonomy to anyone else, especially the state. From the first comes the idea of moral responsibility and therefore the liberal claim that individuals are capable of moral agency; the second denies to the state the right of coercion in moral matters and to anyone other than the individual moral agent. Autonomy makes a certain level of community a requirement of liberal society as the product of tacit (though inevitable) consent but denies to the community the power of moral force over the individual, since such a delegation of moral authority absolutely requires the expressed consent of citizens.

What results from the application of these two basic principles of autonomy and reciprocity is liberalism's communitarian conundrum. By insisting on the autonomous self-definition of the liberal citizen, liberalism denies community in a full sense by denying the possibility that individuals find their real identity in their interactions with others. Liberal community, Sandel admits, may be "instrumentally" necessary to achieve individual interests and ends, but they are not ends "constitutive" either of a truly moral communion or of new individual moral identities:

> But a self thoroughly independent as this rules out any conception of the good (or of the bad) bound up with the possession in the constitutive sense. It rules out the possibility of any attachment (or obsession) able to reach beyond our values and sentiments to engage our identity itself. It rules out the possibility of a public life in which, for good or ill, the identity as well as the interests of the participants could be at stake. And it rules out the possibility that common purposes and ends could inspire more or less expansive self-understandings and so define a community in the constitutive sense, a community describing the subject and not just the objects of shared aspirations.[47]

Sandel's reference to liberal "common purposes" as insufficiently constitutive of the identities either of the community as a whole or of constituent members is aimed at William Galston's attempt to ground liberal community in what he calls "liberal purposes."[48] Galston's well-known argument seeks to answer the charge, made by Sandel, MacIntyre, Charles Taylor, and others, that liberalism is devoid of the kind of unified (and unifying) conception of the "good" that underlies real community by making autonomy subservient to the communal pursuit of that good. The good was supplanted within liberal theory by the "right" as the measure of moral behavior beginning with the philosophy of Kant, communitarians charge; thus, the resultant and oft-invoked "neutrality principle" protects individual liberty and diversity at the expense of communal unity. In its effort not to discriminate (i.e., to remain neutral) against or toward anyone's understanding of the good (or the good life), liberalism, communitarians contend, relinquishes the fertile ground from which community springs. Only a shared and fulsome conception of the good can generate communal identity between and within individuals.

Galston responds with a thesis that liberalism presents a

> distinctive ensemble of public purposes that guide liberal public policy, shape liberal justice, require the practice of liberal virtues, and rest on a liberal public culture. Liberal purposes, so conceived, define what the members of a liberal community must have in common. These purposes are the unity that undergirds liberal diversity; they provide the basis on which *e pluribus unum* ceases to be a raw and shifting balance of contending social forces and becomes instead an ethically meaningful characterization of the liberal state.[49]

Liberal purposes prove that historically the neutrality thesis has not been a part of liberalism's view of social interaction, Galston argues, and has only been presumed to be so by recent liberal thinkers such as John Rawls, Ronald Dworkin, and, especially, Bruce Ackerman.[50] Even though not constituting the perfectionist summum bonum of classical or medieval political thought, the liberal view does "adumbrate a small number of basic goods held to be worthy of special attention in individual and collective deliberation."[51] Galston further argues that this unifying theory of the good has been present in liberal theory since Locke, his famous toleration of religious diversity notwithstanding, and it provides a moral basis on which to build the liberal communal identity.[52]

Galston's argument is an important one and succeeds in relieving liberalism of some of the more extreme accusations of anticommunalism raised by MacIntyre and Roberto Unger. Yet it is also clear that Galston's understanding of liberal community is not the same view of the nature of community as that of these extreme communitarians, or even that of the more measured Sandel. The reason is evident and goes back to Locke. Galston's "liberal purposes" are the product of an acceptance by liberal individuals of the need for reciprocity to achieve their ends. Galston convinces one that these ends are not merely economic interests, as Sandel and the communitarians claim; rather, they are necessary for the well-being of the individual in *every* sense. Locke, after all, recognized that what was at stake in the acceptance of reciprocity was, as he quoted Hooker, "a life fit for the dignity of man."[53] But the moral requirement for reciprocity and therefore for *some* form of community does not extend to the point where it encroaches on the other primary liberal value of autonomy. Because it does not, neither Locke's nor Galston's communal sense is as all-embracing of the individual as that sought by communitarians. The question for us remains: is Locke's and Galston's community sufficiently reciprocal to meet the personal sacrifices demanded by effective policy making for emergent technological risks? The persistence of the NIMBY syndrome supplies empirical evidence for a negative answer.

Community based on reciprocity but not autonomy is community built upon tacit consent. That is, certain communally shared concepts (for Locke) or purposes (for Galston) come to be accepted by citizens as a matter of necessity to meet other needs. They are shared *instrumentalities*, it is true, though by itself this does not make them trivial in their unifying effect. Yet Locke, Galston, and all true liberals are constrained from taking the next step that would build community on the basis of autonomy. Autonomy is too important for individual identity—it cannot be sacrificed if the individual is to possess meaningful liberty. In other words, for all Lockean liberals communal identity is not the true identity of autonomous individuals. Such a "constitutive" communal identity can never be the product of tacit consent, but only of an expressed consent. But this expression can never legitimately be given by liberals—it gives away too much of the liberal store: the autonomy of the individual. By way of comparison, conservative Roberto Unger is willing to take the step that would root individual identity fully in "organic groups," in full awareness of what is being forsaken.[54] This step can only be taken as an act of will, accompanied by an expression of consent; liberals cannot, therefore, consciously consent to take it.

Fully constitutive community is not completely compatible with democracy, then, in the liberal view, because it grants too much power over the individual to the community of the whole.[55] For liberalism to accept this view of community would require guarantees that expressed consent would not forever give away the individual's right to autonomous choices. These guarantees are not present in Locke's theory, nor are they in most other versions of liberalism, since they intrinsically work against social contract theory in particular and representative political institutions in general.[56] Without the guarantees, the community is left with insufficient power to redefine individual life as communal life, that is to say, to change individual identity.

In summary, then, liberalism's dedication to the idea of consent limits community to that achievable through the workings of tacit consent, such as in Locke's example of the introduction of money. Such consent provides more community and grants it more moral significance than liberalism is usually credited by communitarians but does not give rise to, in Sandel's terminology, truly constitutive community. Constitutive community violates the liberal right of expressed consent and thereby compromises the liberal goal of individual autonomy. Although liberal societies may manifest a strong sense of communal belonging to the place one feels most at home—that is, one's hometown or neighborhood where the literalness of being reciprocal parts of something is strongly felt—communal attachment to the abstraction of society as a whole, with its accompanying organic conception of the good, remains weak.

Whatever else its failings, the liberal version of community is insufficient to respond to the risks of modern technology. The paradox of liberal community in the age of technological risk is a direct result of risk's emergent nature. Modern risks emerge from the community-wide choices—tacit and express—that make up the developmental histories first of science and technology and also of technologically developed societies. Furthermore, those risks pose nonseparable consequences for society as a whole, for example, the risks of air pollution or radioactive contamination. In other words, the emergent risks of modern technology are communal themselves in both their sources and their effects.

Liberalism's version of community is not up to the challenge that these risks pose for two reasons. First, due to its reliance on reciprocity rather than legitimately transferred autonomy, liberalism's sense of community is, at best, too parochial to deal with the very national (really, global) risks presented by technology. Second, because of its reluctance to transfer autonomy except on the basis of expressed consent, and the dearth of opportunities to actually express consent when needed, liberalism's institutions lack the legitimacy to exercise the power and command the obligations—and often sacrifices—that a realistic political response to technological risk presumes.

On the first score, contemporary politics is full of disputes between communities concerning how to deal with new risks. Towns compete over the placement of dangerous facilities, and even states go to court against each other to prevent federal policy implementation within their borders. An obvious example of both is in the diversified area of waste disposal, either "merely" chemical or radioactive wastes. Waste disposal is clearly a national policy issue, but its solutions are locally sited. Any solution, then, will require citizens surrounding local sites to demonstrate a sense of community and responsibility that extends beyond their immediate experience in the neighborhood or town, but since community obligations can be based only on reciprocity, reciprocal relations are difficult to recognize, maintain, or even justify on the national scale. Federal compensation programs might help build both a sense of communal sacrifice and reciprocal indebtedness but are as yet largely absent.

Liberalism's second communal failing regarding risk policy arises from its value of autonomy. Solving national risk problems will require the exercise of power of a kind and to a degree that makes liberals nervous and fear for the continued safety of the value of autonomy. Siting issues often invoke the communal right of eminent domain, but its exercise just as often fails amid the clamor of public disputes about individual rights of property and other aspects of autonomy. It is not difficult to envision institutional procedures that might temporarily transfer to society the power sufficient to take required action; that is, temporary grants of power to which expressed consent is given. But such processes of what Amy Gutmann and others call "deliberative democracy" are not common in liberal societies, at least at the national level, and representative institutions are loathe to allow them entry into the domain of political authority.[57] Without such participatory measures, however, expressed consent cannot be presumed or usually even considered. Thus, the power to make the tough risk decisions lacks legitimacy.

Conclusion: Risk and Liberal Identity

Identity, says William Connolly, "requires difference in order to be, and it converts difference into otherness in order to secure its own self-certainty."[58] Thus, the certainty of self-identity is always in ontological tension with the otherness of community but never more so than within a liberal culture that defines itself using concepts that stress the reductionist instead of the emergent aspects of social behavior, the individual and the discrete rather than the collective and the common. Within such a context, coping with risks that themselves cannot be broken down into discrete and individual aspects is difficult, even when there is a pressing need to do so. Within a culture that views risk as the privately chosen consequence of individual decision and action, the ideas of either communal risk or communal solution are negligibly compelling.

Yet all current models of risk assessment presume that individuals make decisions about community-based (i.e., emergent) risks and that in so doing they give their consent to policies generated to cope with the risks. Furthermore, these approaches insist that the consent need only be tacit, not expressed, even though a considerable degree of individual autonomy over the choice of risks is being transferred somewhere else, usually to a governmental organ itself claiming only the legitimacy of tacit consent. Is it any wonder that these approaches to risk assessment and policy making have resulted in NIMBY claims and litigation that slowly suck the life out of the policy process itself, while at the same time raising the essential political question of whether liberal democracy is capable of finding solutions to problems raised by the technologies and science to which it is so wedded?

A more compelling sense of community, and a fuller notion of moral responsibility, is the answer to the political question manifested in the NIMBY phenomenon. Liberalism has probably always known this, but denied the knowledge on the grounds that such an answer would forever consort with the enemies of individual liberty and of the right of the governed to consent. Therefore, Locke allowed that tacit consent could generate whatever moral community was really necessary, but that beyond a loose collection of individuals pursuing their (perhaps joint) liberal purposes, expressed consent would be required. The right of autonomy disallowed the granting of expressed consent, however, for no political obligation could ever legitimate the exercise of force against individuals sufficient to make them acknowledge a moral obligation sacrificing their own interests to those of the community.

Modern risks and the policy dilemmas they generate illuminate the essential outdatedness of liberal understandings of political obligation, authority, and community. Modern risks bring us ever closer in our jeopardy and ever nearer the realization that since risks emerge from our collective identity, so must their antidotes. But seventeenth-century ideas reducing all political phenomena to atomistic components remain as hurdles over which effective risk policy must leap. Doing so will require a different understanding of obligation and authority, of the meaning of democracy, and of individual identity. This process of generating new definitions portends, as Connolly notes, a more prominent role for the exercise of power; thus, the issue of authority is central to a more effective politics of risk.[59] These new conceptualizations bring individual citizens to a new beginning point in the constitution of their political identities, at the point of engagement with rather than detachment from the community, where property lines blur and fences disappear, making it difficult to shout "not in my backyard."

Risk and the Authority of Ends

Within liberal theory it is clear that consent is an important—perhaps the only legitimate—basis of political obligation, in addition to providing, as in Locke's theory, a significant source of moral obligations. Obligations, in turn, always establish authority of some variety, if only in the minimal sense of "authorizing" a set of expectations as legitimate for some person or persons to hold as a consequence of our having entailed an obligation. If, as we saw in chapter two, the obligation is moral, the actual authority of the persons (or community) to whom we are obliged does not necessarily entitle them to do any more than attempt to persuade us to live up to our obligations or possibly to shun us if we do not. If, on the other hand, the obligation is legitimately construed as political obligation, consent grants a fuller range of legitimate authoritative responses. As we have seen, liberalism's troubles with political obligation and authority are not really solvable by its construal of consent, because such consent would need to be explicit and expressed, yet liberal societies offer few if any opportunities for acts of expressed consent.[1]

As introduced in the last chapter, liberal theory—both classical and modern—presents two equations of consent: the first giving rise to society as a whole and its structures of collective decision making and moral community; the second form of consent to persons who occupy those structures. William Connolly says, "Consent is in fact necessary to the modern practice of authority, but consent, when it is offered, applies not only to procedures for decision but to a larger set of understandings, ends and purposes that help to constitute a way of life."[2] As shown in the last chapter, modern risks challenge the consensual foundations of the liberal community by exposing the weakness and insufficiency of the liberal obligation to collective authority. Risk policy requires more than the authority of moral suasion, as most NIMBY disputes make clear. It demands the power-laden and legitimate authority that within liberalism comes from political obligation. Such authority is exercised by authorized persons within offices presumed to be legitimate, and it is assumed that such authority will be exercised with sufficient vigor to ensure whatever level of compliance is necessary to protect the public good. This authority is the product of the second liberal equation of consent—consenting to particular persons or offices designed to carry out the collective will of society. This chapter explores how modern technological risks challenge this notion of consent as well as the authority to which it gives rise.

Modern political science begins with the belief that when citizens grant their consent and assume a political obligation, it is with the understanding that the recipient of their trust now is entitled to exercise authority. The previous chapter elaborated how this seemingly axiomatic bit of political science lore brims with ambiguity. Some of the uncertainty of consent and obligation has to do with the establishment of moral obligations between individuals within the collectivity of modern society. The remaining elements of this hoary political equation—trust and authority—are the focus of this chapter, and they present their own conceptual difficulties as well as a problematic empiricism.

Defining, much less locating or recognizing, authority within the liberal state is never easy, since, as Sheldon Wolin remarks, liberalism has always conceived of society "as a network of activities carried on by actors who knew no principle of authority" and are therefore "untroubled by [its] presence."[3] Our task becomes more difficult within the often highly charged and raucous political context of technological risk and its accompanying public disputes when citizens become *very* troubled by authority's presence. This chapter argues that within such public debates and protestations a new notion of authority is being forged, one bearing little likeness to that envisioned by many liberal political theorists living during liberalism's birthing in the seventeenth century or during its maturation in the twentieth. It is one that severely challenges the liberal notion of authority and its basis in consent.

As a general rule, the perceived presence of risk challenges existent ideas and states of affairs. This is one reason why, in politics and policy making at least, perceptions of risk are so important, although even unrecognized risks can have political significance once their effects have become manifest. In a political context, technological risk often calls into question current practices or actors within the decision-making process, regardless of whether citizens view the process itself as the product of consent or the actors themselves as legitimate officeholders. In cases like these, the grounds for citizens' dissatisfaction with their leaders (or the specialized subset of them sometimes called "risk professionals")[4] are the anticipated consequences of the latter's decisions. That is, the negative consequences of living with a risk (what I will later call "ends") serve to delegitimate the authority of the actor or process that makes the policy likely to deliver those ends. In other words, authority loses the trust of those it is empowered to command. For instance, during the early 1980s policy struggles to confront the risks posed by problems of radioactive waste disposal, the US Office of Technology Assessment concluded that "the greatest single obstacle that a successful waste management program must overcome is the severe erosion of public confidence in the Federal Government."[5]

Radwaste disposal policy is not the sole reason behind the ebbing of public trust in government, of course, although we will return to it later as an example of the general phenomenon of the decay of authority in liberal societies. But this and other risk policy areas are contexts in which the reasons behind this political phenomenon show themselves. Because trust in the ability of government and decision makers to govern well is an intrinsic feature of the decision to consent, decline in that trust further weakens the allegedly consensual basis of modern politics and authority.[6] "Governing well" here depends, according to Joseph Raz, on the ability actually to make decisions and solve problems without violating the community's standards of correct moral conduct. This Raz calls the "dependence thesis," and it is an essential requisite of the legitimation of authority in contemporary liberal societies.[7]

Not surprisingly, declining trust in government's ability to make difficult decisions is directly related to the emergent nature of such risk issues. Since the risks emerge from the fact of social life and are often not obviously matters of either individual choice or collective decision, any governmental action is likely to be seen as (and be) reactive rather than anticipatory. This makes policy making difficult for a variety of reasons, but especially because institutional patterns of authority and delegation do not necessarily match the policy demands of emergent technological risk. Don K. Price begins his exploration of the "unwritten constitution" of the United States by commenting on the "confused sources of authority" within the American science and technology policy system:

> We may often find that there is no center of political authority in existence with the capacity to put into effect a rational comprehensive policy. The United States government is not a single rational actor. The degree of independence of congressional committees and their staffs and the related independence of specialized executive bureaus and agencies make it impossible to plan and carry out a coherent program. There is no clear center of authority in either the Congress or the executive that the voters can hold responsible.[8]

Price's fears about the absence of authority over risk policy, as those of one of the first chroniclers of the rise of the science policy establishment in the United States, are worth exploring further, especially because, as we will see, his sentiments are echoed by many others as well. I begin this chapter with a broader focus, however. Technological risk not only exposes the core absence of authority within the policy-making centers of government; it also raises fundamental issues about the proper source or basis of authority to make those decisions. Authority in liberal states, as Price's focus on the United States provides an example, has never been completely secure or rooted in any one location, argument, or tradition. Liberal citizens recoil at their self-image as subjects, Price argues, and have since colonial days "sought to escape even responsible authority by various automatic formulas for salvation or for secular perfectibility."[9] Such escape attempts plausibly explain the current cachet of some recent popular proposals for constitutional change in the US, for instance term limits for legislators (certainly an "automatic" mechanism), but, in the case of term limits as well as in risk policy, these mechanisms do little to alter the authority vacuum at the heart of many policy issues.

Liberal societies, in short, not only have a policy problem with authority; they have a more basic and widespread ambivalence about authority that is subsequently reflected in the policy realm. Technological risks and policies for them call attention to this ambivalence and, importantly, exacerbate it, to the further detriment of authoritative decision making. To see how this is so, we must begin at a general level of analysis about the concept of authority within political theory and politics. Thus, the first section of what follows briefly investigates the meaning of authority in its historical development as a political concept. Also in the first section, I discuss the contemporary meaning and inner logic of the authority relation.

The second part of this chapter explores how liberalism as a theory and organizing principle for liberal society has altered the meaning of authority and separated it from its classical roots. Here we will see how liberalism "divides, disciplines, and defers" authority within liberal society, to the general effect of virtually eradicating it from contemporary politics.

In the third section I explore the impact on liberal authority of technological risk and the policy demands that arise from it. Here we will see how risk politics make it clear that the liberal construal of authority is simply not up to the challenges that modern risks pose for decision makers. Today this is most visible in the reliance on courts to render decisions that effectively become whatever passes for risk policy in the United States. Since judicial authority has little chance of surviving the strictures of Raz's "dependence thesis," the debilitating impact of risk on liberal notions of authority becomes clear. Finally, in this chapter's concluding section, I propose a different conceptualization of authority in liberal societies confronted by risk, one that holds some promise as an effective decision-making device within the politics of technological risk. I call it the "authority of ends."

Authority in Politics

Compared to other central concepts in political theory and politics—for instance, justice, rights, or liberty—authority has received considerably less scholarly attention in recent years and arguably since the days of Aquinas. Partly this is due to liberalism itself, or even more broadly to modernity, with its beginnings in Machiavelli's dark exploration into the achievement and exercise of power and its alleged end in Nietzsche's anticipation of the postmodern nexus of knowledge/power. The relationship between power and authority must be clear at the beginning of any discussion of the latter, since the facile "authority is legitimate power" ignores what Machiavelli and Nietzsche both publicized—that power can and does legitimate itself merely by its successful exercise.[10]

Modern liberals, possibly excepting Hobbes, certainly did not emphasize the concept of authority, seeking instead to ground politics upon the concept of individual interest and the priority of the right (defined institutionally) over the good (to be defined by each individual). It was the latter idea of political "goodness" that animated most medieval and classical conceptions of authority, from Plato's icon of the philosopher-king to Aquinas's sanctified secular leader doing the work of the Almighty (under the direction of his vicar, the Pope). Authority in such conceptions carries an ontological commitment stressing that just as true goodness exists outside the individual (and is, therefore, never defined by individuals, but by the community), so then must authority, since its foundation lies in the knowledge of the good. Images of authority as rooted in the community that are reminiscent of Plato's or Aquinas's have only occasionally been resuscitated during the modern era. Rousseau's Legislator claims an authority most reminiscent (to Rousseau also) of Moses, and contemporary communitarian Alasdair MacIntyre wistfully advocates a return to the medieval understanding of the sanctified community serving a divinely ordained authority and brandishing its own authority grounded in moral obligation.

Most modern renderings of the idea of authority begin, however, with an entirely different ontological commitment. Since within modern political thought goodness is defined privately as a matter of individual autonomy, authority at its base belongs to all individuals and can be transferred only with their consent. But even when delegated, authority is not attached to other individuals due to their special insight into some externally existing good, but simply becomes the property of an official institution or office. With an ontological basis rooted in subjectivity, then, modern authority, according to Connolly, is both "fragile" and "ambiguous."[11] Another way of characterizing modern authority is recognizing its investiture within institutional settings, where it is exercised according to a set of principles agreed upon beforehand by citizens who consider themselves singly as equal to each other and to anyone exercising authority. In other words, modern notions of authority begin with the ideas of equality and constitutionalism.[12] Both beliefs serve to limit authority and to secure it within well-defined public positions, usually filled by alleged acts of consent (such as voting), performed by the governed. As described by the authors of *The Federalist Papers* and piquantly critiqued by de Tocqueville, this institutionalized notion of being "in" authority is for some classically minded theorists like MacIntyre, by way of contrast, a denial of the concept of authority itself.

Hannah Arendt was one of the first to maintain that the modern notion of "in" authority "not only challenged one or another form of authority in different spheres of life but caused the whole concept of authority to lose its validity altogether."[13] For Arendt the subjectivist ontology of modern views of authority—both liberal and otherwise—were realistically a denial of the true ontology of authority as existing "outside the range of human deeds."[14] In her view, the loss of the true concept of authority precipitates a crisis that threatens to replace a socially rooted and meaningful human existence with "mass-society," a "kind of organized living which automatically establishes itself among human beings who are still related to one another but have lost the world once common to all of them."[15] This is the crisis of authority, and what it portends, Arendt claims, is a loss of all ontological certainty:

> Authority, resting on a foundation in the past as its unshaken cornerstone, gave the world the permanence and durability which human beings need precisely because they are mortals—the most unstable and futile beings we know of. Its loss is tantamount to the loss of the groundwork of the world . . . the loss of wordly permanence and reliability—which politically is identical with the loss of authority.[16]

As the antidote to mass society, authority for Arendt starts with the idea of shared understandings that authoritatively order the expanded political world of individuals and define their place in it. This is a distinctly unmodern view of authority and not liberal either, because it does not leave those understandings to the varied and autonomous definitions of individual citizens. The shared beliefs and values that constitute authority for Arendt, in other words, are more than merely the shared interests of individuals; they are traditions carrying what the Romans called "auctores imperii Ramani Conditoresque—the authority of the founders."[17] They provide a realm of meaning constitutive of a life together and of each person's place within that life.

Authority for Arendt is both product and species of truth and, as such, is not reducible either to power or even persuasion. Authority presumes obedience, it is true, but not through threat of physical force or even the force of argument. Authority is an "augmentation" to decisions, of a nature both elusive and intangible, which, quoting Mommsen, she describes is "more than advice and less than a command, an advice which one may not safely ignore."[18]

But what exactly is added to decisions that makes them authoritative, not merely coercive? This issue goes to the heart of the distinction between classical and modern notions of authority. For Arendt, the augmentation that authority attaches to decisions derives from the history and tradition of a community that sees itself as a moral entity. Thus, the Greek polis carried the authority of knowing not only about the past or about the physical world but about how to live one's life. Authority in the classical view, then, adds a normative presumptiveness or expectation of obedience that comes from the bonds that unite individuals into a shared life.[19]

The idea that authority presumes shared values is common even among contemporary theorists of the concept. Though adopting the modern liberal stance that even "an" (or "epistemocratic")[20] authority resides only "in" institutions, Richard B. Friedman sees the distinction between "an" and "in" authority as lying in the presence or absence of these shared values. "An" authority is present in institutions

> to the extent to which the members of society regard those institutions as reflecting, embodying, or promoting their shared beliefs, and the dominant theme of this approach is that, although a common set of authoritative beliefs is constitutive of the social order, the weakening or dissolution of those beliefs is bound to generate destructive acts directed against the values and practices of the established social order and even ultimately against the self, e.g., suicide, madness. It is usually some version of this approach to authority—in terms of shared beliefs—that those writers have had in mind who claim that the notion of authority has been lost or distorted in modern times.

"In" authority, by contrast, is inextricably linked to the modern, particularly the liberal version of community, which, as we saw in the last chapter, is not based on the sharing of core values of goodness or moral virtue. Thus, for Friedman, "in" authority

> is political authority that is needed just because men are "individuals" who do *not* share the same values or, what is the same thing, who conceive themselves to have different and often conflicting purposes in life, so that no social order whatever could be maintained among such individuals if each were to insist that government reflected his "values," as a condition of acknowledging the duty of obedience.[21]

In Friedman's typology, it is clear that Arendt perceives traditional authority as a moral authority carrying the forces both of objective truth and of a shared tradition and therefore needing no additional warrant from individuals for their obedience. The reliance on a strong sense of community is a major part of this version of authority and implies that as the communal sense wanes so too atrophies the whole structure of authority. Wistful for times past when authority manifested social rootedness, Alasdair MacIntyre agrees with Arendt's presumption that authority must carry a notion of moral certainty for it to qualify:

> In our society the notion of moral authority is no longer a viable one. For the notion of authority can only find application in a community and in areas of life in which there is an agreed way of doing things according to accepted rules. There being an agreed right way of doing things is logically prior to the acceptance of authority as to how to do things. . . . [U]nless there is an established and shared right way of doing things, so that we have social agreement on how to follow the rules and how to legislate about them, the notion of authority in morals is empty.[22]

A central aspect of the classical view of authority—and one accepted by Arendt and MacIntyre—is that a shared set of values and beliefs means that "an" authority presupposes some kind and degree of suspension of will or blind obedience by those subject to it. Henri Bergson refers to the "unthinking submission" such authority implies;[23] Arendt calls it "unquestioning recognition"; Friedman notes the "surrender of private judgment." And since this is a form of moral authority being exercised, "an" authority here is entitled to expect, in Elizabeth Anscombe's plain language, that subjects are "liable to" obey its dictates.[24] In short, this is authority requiring no further justification than that which it automatically possesses as the product of historically shared values still in force. Persons subject to this authority are not prevented from demanding justification, Friedman points out. Rather, it never occurs to them to seek it.[25]

It is with the classicists' surrender of private judgment as a feature of authority that modern thought departs most obviously from the classical view. This is most clear within liberal approaches to authority, which, beginning with Locke, deny the external ontology of authority in the classical view and replace it with the notion of consent as that which legitimates authority claims. Thus, the locus of authority is moved by Locke from the community *external* to the individual to the consent emanating from *within* the individual. In so doing, liberal theories of authority reject the requirement that individual judgment be suppressed or replaced. The operative value behind this shift is obviously individual autonomy, which, as the guiding value of liberal theory, guarantees that no external authority can legitimately coerce individuals into forsaking their own judgments concerning the good life. The strongest contemporary rejection of the classical view for its denial of individual autonomy remains Robert Paul Wolff's *In Defense of Anarchy*; less extreme, though kindred positions are those held by Raz, Rosenblum, Flathman, and many others.[26] As we will see, risk politics presents a hybrid version of both classical and liberal views of authority. The ubiquitous presence of technological experts has resuscitated the classical foundations of "an" authority (seemingly without the normative claims) as lying in the specialized knowledge they hold, while at the same time NIMBY claims obviously adopt an extremely liberal (or is it Wolff's anarchistic?) view.

Still, from this brief characterization of the classical view's normative vision of authority, it is evident that Arendt and MacIntyre deliver timely eulogies to it. The killing was presided over (if not actually committed) by Machiavelli, who, in his concern and contempt for papal power in Italy, sought to replace the church's undeniable moral authority with a new model based not on shared values but on the vicissitudes of power and interest.[27] Consequently, he changed for the next 400 years the meaning of political authority, locating it inexorably within institutions as diverse as Hobbes's Sovereign or the US Constitution's three branches of government. Authority had become irretrievably attached to institutions ("in" authority) and thereby essentially detached either from the shared values of the community concerning the "good" or from their personification in the person of "an" authority.

One of the many ironies of Machiavelli's political vision is the oddness of its debt to the classics, especially to Plato. Although Plato's philosopher-king and Machiavelli's prince seem antithetical, both share some important attributes of their claim to authority. Just as the philosopher-king patterns the divisions of the just society to accord with those within his own soul, so Machiavelli has the Prince fashion the modern state in his own self-assured and autonomous image. Both philosopher-king and prince operate in their own rarefied and self-legitimating realms of political necessity, and both are chiefly concerned with laying the foundations of the political community. But the political personae of the philosopher-king and the Prince are wildly at odds, and so the foundations they lay are in altogether different ground.

In *The Republic*, Socrates argues initially with Thrasymachos that the foundation of the philosopher-king's authority lies not in his own interest but in the interest of his charges, much like the cases of a shepherd, physician, or ship's pilot. Yet Socrates does not go on to build a case for democracy, for the interest of the many is not really the concern of the political community, any more than is the interest of the philosopher-king. Rather, the polis exists to participate in the eternal order of justice, and only the philosopher-king can glimpse this ontological, as well as moral, truth and what it indicates for political organization. Furthermore, the whole issue of what power the philosopher-king wields or even the distribution of power within the republic is never really Plato's major concern. The power of justice is separate from the concerns of men, and what men mean by power serves not the interest of justice but only their own interest. To serve within the realm of justice necessitates that politics begins with the idea of duty, not power or interest; and to realize this noumenal realm in the world of phenomena is to harmonize all duties within the hierarchy of the just soul. Physical force can never achieve this harmony or this justice; thus, the authority of the philosopher-king is of a different—and for us moderns, as Arendt mourned—largely unrecognizable sort.

For Machiavelli, the beginning of the civic principality lies not in duties but in "two opposing interests," a vision reminiscent of Thrasymachos and encompassing a view of power all too recognizable today. "In every city," Machiavelli writes, "these two opposite parties are to be found, arising from the desire of the populace to avoid the oppression of the great, and the desire of the great to command and oppress the people."[28] Only the overwhelming physical power and presence of the Prince can achieve a synthesis of these two conflicting interests, and at his disposal must be his own "cunning assisted by fortune."

Three elements of Machiavelli's picture of the Prince's authority distinguish it from Plato's and make it modern. First, the authority of the Prince has more a sense of place than of person. It is authority that has a residence somewhere in the world of phenomena, in an institutionalized office, filled by a very real person with specific and (by some standards) not altogether laudatory attributes. The power of authority exists—it is *there*; what politics awaits is the right person to take it up. The Prince does not bring the power with him—it is not his merely through the force of his personality. It belongs to fortuna; it resides in the collective world of politics.

Second, and already implied, the Prince's authority is reducible to the power he wields, not the knowledge he attains about value or shares with others. There is no legitimacy for authority other than that power makes it so. Machiavelli begins the process orientation of politics and thereby replaces the classical fixation on ends. Unlike the classics, Machiavelli refuses to see the reason that moves the Prince to grasp power as some "preordained end or purpose of human existence," embodied only in the politics of virtue. Rather, reason ought to be seen as the "marvelously effective *servant* of the passions," passions that with their instrument of reason bring to the political realm the possibility of diverse political forms.[29] True, a case often is made that Machiavelli legitimates the use of power by its achievement of the nation state, but even if that is a valid interpretation, the ends of the state are of little interest beyond the fact of the unification of power. Sufficient power *is* authority, and its ends are a matter of interest only to those who exercise it; this is true whether the Prince is viewed as a monarch or, as Machiavelli notes in the *Discourses on Livy*, in a republic as the whole people. If they (or he) exercise power successfully, the ends that authority serves need no other justification.

Third is the realization that, for Machiavelli, interest is indeed the stuff of politics, the only variable that "fortuna" does not affect. Politics happens because human interests conflict, and therefore resolutions are necessary. *The Prince* as a book is a chronicle of what different institutional resolutions have historically been chosen, of which ones have worked better than others, and of what they required. Politics is made necessary by the inescapable fact of conflicting interests, and Machiavelli is himself ambivalent about the institutional context of the resolution or what it portends for the structuring of the political realm. It does not matter, in other words, if he is correctly interpreted as a republican seeking to subvert autocracy or as an advocate of a most pernicious autocrat in Cesare Borgia. Machiavelli offers the opportunity to choose autonomously our institutions of authority, because ontologically they concern only our own interests, not eternal truths about existence or the good life. We have the power to create politics since politics is *for us*, for our interests. Rather than the eventual degeneration of Plato's Republic, we have the "Machiavellian moment" of choice and creative power sufficient to establish the state. Tom Paine said it for all of modernity: "We have it within our power to begin the world anew." It is the only authority the modern citizen needs.

Authority in Liberal Politics

The authority of Machiavelli's prince derives from his role as founder, but not a founder as in Plato's vision of the philosopher-king or in the Roman sense of "auctor." The Prince possesses no special insight or knowledge into the "good;" rather, he manifests a new institutional order within which authority is located. Machiavelli is not out to identify a realm of meaning in politics as much as a realm of order, power, and interest. The politics that results is a process largely without any defined collective goal or "end," and whatever sense of authority it maintains is identified with a step or office within the political process, with purposes entirely secular and nonmoral. The concept of authority is truly denuded of its classical referents denoting tradition and shared meanings constitutive of a community serving an external purpose or end.

Liberalism completes this dispossession of authority by adding elements to the political process that remove any remaining vestige of authority's larger classical role. These are elements geared further to limit the operation of "an" authority and to cement a form of "in" authority firmly within offices not defined—as arguably they are in Machiavelli—by personality or character. Liberalism, then, clearly takes Machiavelli one step further in the modern reconfiguration of authority, attempting to fully depersonalize authority by attaching it to positions rather than to persons.

Following the Machiavellian conflation of authority with power, liberalism adds three procedural elements to the princely power, all of which begin with the new subjectivist ontology for authority grounded in consent. They are first, dispersal; second, what George Kateb calls "chastening;" and third, combination with expertise. In institutional terms, these functions are recognizable as residing first, within the separation and division of powers and its resultant bureaucracy; second, within representative government; and third, within what Don K. Price named two decades ago the "scientific estate." Together, these elements are meant to accomplish the liberal limitation of power. In reality, they accomplish only the burial of authority and its replacement with the pallid protectionism of the postmodern state, the insufficiency of which becomes clear in the systematic weighing and rejection of competing versions of authority evident in risk policy decisions and disputes. It is in these disputes surrounding risk that the modern bargain with authority is unmasked as more Faustian than Machiavellian and where giving the devil his due motivates citizens to reject the liberal conception of authority altogether.

Authority's Liberal Diaspora: Dividing, Disciplining, and Deferring

In its fervor to protect the individual from unwarranted authoritarian interference, liberalism as a political doctrine is a great engine of institution building. The first creation is of course a constitution, which in the case of the United States is dedicated to erecting and delimiting the functions of several other institutional constructs. Not all liberal constitutions follow the mechanical model of the United States, with its careful calibration of checks and balances of power. Nevertheless, it is basic to liberal theory that institutions are there to protect individuals from the arbitrary (read "individual" or "personal") exercise of power, and the more mechanistic the protection, the less reliance on the overly personal and vaguely threatening forms of "an" authority.

The US Constitution of course disperses authority along both vertical and horizontal axes, that is, between the federal government and the states and between the three branches of both levels of government. But the liberal investiture of authority into visible offices with clear job definitions does not end with constitutional conventions, and therein lies the cause of further dispersal of authority. The United

States constitution is, as is well known, extremely vague in the granting of authority and filled with empowering ambiguities such as the Tenth Amendment and clauses like "necessary and proper" and "due process of law." These features may well be largely responsible for the Constitution's record staying power, but they also lead to the erection of new sources of authority not recognized by the Constitution. The gradual accretion of power by the executive branch and by the national government as a whole throughout the nation's history was perhaps predictable as the nation itself rose in world prominence. But the accumulation of power in Washington, DC, did not dwindle the focal points of authority; rather, it multiplied them.

Because authority in liberal governments and societies refers only to an interest ascendant in the exercise of power, it is not difficult to understand how the proliferation of points of authority came about. As power became ever more centralized in the institutions of national government, the application and administration of power required new organs of authority. Thus, the bureaucracy of the executive branch fanned out across the political landscape, and the parallel congressional committee system replete with its legion of staffers took hold. The power wielded by such instrumentalities made the authority of elective office even more attractive; then the elaborate superstructure of the electoral system grew important, with its parties, conventions, primaries, as well as the tools necessary to win the day: media, political handlers, consultants. Each of these represents an important locus of political power and influence, yet none are mentioned in the Constitution. They embody what Price identified as the "unwritten constitution":

> —the fixed political customs that have developed without formal Constitutional amendment, but that have been authorized by statute or frozen at least temporarily, in tradition. The party conventions and the primary system, congressional committees and their staffs, the statutory structure of executive departments, the Executive Office of the President, the press conference and television coverage, and freedom of information—none of these was established by the Constitution or foreseen by its framers, and all could be abolished without amendment.[30]

For Price, the unwritten constitution invests authority for policy making across a vast field of actors, only some of whom are constitutionally warranted to embrace it and concerning whom the public may remain largely unaware. Centers of power arise because interests are engaged, and without specific constitutional legitimation, such centers exercise authority over important policy issues. Price's example is a common political science one—the existence of "iron triangles" made up of lobbyists, bureaucratic agency chiefs, and congressional committee chairs. The power such allied forces wield is prodigious, Price observes, because of their uniquely modern wedding of authority and interest rather than of authority and value. Voters indeed delegate authority to their elected representatives, but it is not to be exercised on the basis of the representative's own values. Instead, Price says that authority rests partly on legal definition embodied in congressional mandates and partly on the technical advice and information given committees by lobbyists and administrative agencies:

> But the strength of the "iron triangle" does not depend on organization and procedure. It rests more fundamentally on what American politicians and the public gen-

> erally believe in. They do not believe in letting political leaders have the authority to make important decisions on the basis of their value judgments. They prefer an approach that is partly scientistic and partly legalistic.[31]

This acceptance of authority strictly as "in" authority on the part of American citizens leads, as Price's argument makes clear, to an ever wider dispersal of authority within American politics. Furthermore, although Price's analysis largely ignores the impact of citizens themselves in the dynamic of authority's dispersion, his argument can be extended to cases in which active citizen involvement in the policy process serves to further the phenomenon. In siting disputes involving the distribution of risk, for example, when citizens have increasingly involved themselves and manifested the NIMBY mentality, the general effect has been even wider dispersal of decision-making authority as more officials dependent upon the goodwill of citizens seek to involve themselves in the dispute. James Thurber recounts how this accentuation occurred around the Nuclear Waste Policy Act of 1982, legislation seeking to identify sites for nuclear waste dumps. Since citizens' action focused (in good liberal fashion) on their self-interest in avoiding proximity to dumps, rather than on the value-oriented issue of social responsibility within a nuclear-powered society, the effect of the legislation was to involve more congressional committees in the siting process as members' constituents demanded protection from their representatives, as Thurber explains:

> Each element of the Nuclear Waste Policy Act raised new issues of significant concern to Congress, state and local governments, and citizens. Members of Congress and their constituents became more interested in DOE nuclear waste policy as potential states were identified for site characterization. As constituent pressure against the siting of the nuclear waste sites increased, so did congressional oversight and the number of committees and subcommittees having a stake in radioactive waste management.[32]

The end result of this proliferation of nuclear policy authorities within Congress (and in combination with their "triangular" counterparts) was as Price predicted: stalemate and the incapacitation of government to make policy. Authority for nuclear waste disposal policy has been dispersed into oblivion.

Disciplining Authority Through Representation

Liberal citizens do, of course, acknowledge the authority of their elected leaders to represent their interests, if not to stipulate their values, in policy issues like waste disposal. Such representation embodies a grant of authority, of course, but only in the sense of an authorization to further the citizens' interest, not an ordination to lead citizens in an exploration of what those interests are or should be. Liberal citizens reserve to themselves the right to decide where their interests lie, and their liberty guarantees that they need defer to no one in the determination of where their interests *should* lie. Least of all need they defer to political leaders for such ethical direction, for within liberalism elective political authority is an investiture of agency, not a consecration of moral leadership.

George Kateb identifies the essential fact of representative institutions within liberalism and liberal politics—they regularly recreate authority and in doing so constantly reassert authority's artificial nature. Although political authority is itself a necessity of social life, representative institutions create an impression that authority is impossible without the willing participation of citizens. When authority can be withdrawn from any of its holders at any time, its nature is fundamentally transformed in its operation, even though its necessity remains undeniable. Representative institutions, Kateb concludes, "chasten" political authority:

> [R]epresentative democracy signifies a radical chastening of political authority. When political authority is, at every moment, a temporary and conditional grant, regularly revocable; when suffrage establishes the sufferance, so to speak, in which the people hold political authority, a major moral distinctiveness enters the life of society. Society is taught—society teaches itself—a fundamental lesson about the nature of all authority by handling the problem of political authority as it does. . . . [W]e speak of a pervasive skepticism toward authority, a reluctance to defer, a conviction that those who wield authority must themselves be skeptical toward their roles and themselves and that necessary authority must be wielded in a way that inflicts a minimum damage on the moral equality of all people. Furthermore, there is a tendency to try to do without authority wherever possible or to disperse or disguise it, and thus to soften it.[33]

Kateb's concept of chastened authority clearly fits with Price's understanding of liberalism's penchant to disperse authority. Both display liberalism's attempt to limit authority's realm to that of agency for the people's interest, and, especially in Kateb's construction, the people are viewed reductionistically as individuals "less fearful of all authority." Such individuals manifest *autonomy*, says Kateb, "the positive expression of independence in the face of personal and impersonal authority."[34] Autonomy is a claim announcing that one cannot be legislated for, that one can only legislate for oneself, and, says Kateb, this is an intrinsically antipolitical claim. Although Rousseau tried to bring this notion of autonomy into line with actual legislation, the homogenization of individual life and the infrequency and simplicity of legislation that resulted could not sustain any real political society. Therefore, in actual political life, the liberal goal of autonomy chastens political authority by severely limiting authority's relevance within a separate public realm. So arises the liberal hallmark of limited government, the benign protector of Kant's autonomous individuals existing primarily within a private realm as "ends in themselves," largely exempt from public authority.

As merely the agent of the people's interest, chastened political authority is not fully trustworthy authority. Representatives are "delegates" not Burkean "trustees;" they are "in" not "an" authority. They are agents of their citizens' interest, not moral arbiters of what that interest should be, and citizens reserve always the right to deny authority, to say no to their representatives. This, as Kateb points out, is the negative side of autonomy, the disposition "to dissent, to engage in acts of principled or conscientious disobedience or resistance or rebelliousness, whether in acts of citizenship or in the rest of life."[35] Together this lack of trust in authority—or better, this lack of investment in authority—and this predisposition toward autonomy make the NIMBY syndrome seem an altogether expected, even defining feature of the liberal reaction to authority.

Most philosophers follow Raz in assuming that modern authority always depends on citizens' trust in the ability of those with authority to make wise—or at least informed—decisions that also do not violate community standards of relevant moral issues such as fairness, justice, or individual rights. Similarly, most observers of the policy conundrums surrounding risk note the pervasive loss of public trust either in the ability or ethical evenhandedness of authority figures to make correct decisions. Sometimes this withdrawal of trust is aimed at a single agency such as the Department of Energy (DOE) or the Nuclear Regulatory Commission (NRC); sometimes, as Pijawka and Mushakatel observe, it "extends to almost all administrators who have a distinct stewardship function."[36] In either case, such loss of trust is a direct result of the dividing and chastening dimensions of the liberal approach to authority. This loss has predictably negative policy consequences, and also some unexpected behavioral ones in liberal citizens. Surprisingly, perhaps, it leads liberal citizens to seek to regain some semblance of classical authority by "repersonalizing" authority—by trying to resurrect some version of "an" (or "epistemocratic") authority. The first beneficiary of this attempt is science and its practitioners; the second is citizens themselves.

Deferring Authority Through Science

The NIMBY response from citizens itself often erects another kind of authority, one that initially seems at odds with the mechanical, representational, "in" authority of liberal theory. Specifically, it is authority often invoked by citizens in opposition to the authority of governmental agents. It is the authority of science, and it is as much a part of the American version of liberal authority as it seems a contradiction of it. Nancy L. Rosenblum argues that

> when political philosophers write about "an authority" today, professional or scientific experts come to mind before religious or moral ones. And here it is fair to say that authority does mean "surrendering judgment." Here, the liberal dynamic of trust, distrust, and judgment is irrelevant. Deference to the superior knowledge of experts is just that, and stems from both a sense of inferiority and faith. Despite increased moves to seek "second opinions," the authority of scientific experts is often taken as a matter of faith in a way that political authority is not.[37]

In a similar vein, Price argues that reliance on science as "the authoritative source of truth" is a prominent feature of the "unwritten constitution" and the only aspect of liberal authority that comes close to the classical sense of "an" authority:

> [S]cience was the reliable source of knowledge and the engine of progress, even if it had to be supplemented by value judgments during our stage of imperfect comprehension. From the days of Thomas Jefferson and John Adams and Benjamin Franklin to President Hoover's Committee on Social Trends and Vannevar Bush's *Science, the Endless Frontier*, even the lay public with little comprehension of scientific method was not disposed to quarrel with the assumption that science was leading us in the right direction.[38]

Liberal society has a long history of looking to science as its only acceptable form of epistemocratic or "an" authority. Indeed, the risks posed by twentieth-century science and technological advance are themselves fruits of an historically intimate association of science and liberal ideas. As a political theory, liberalism has its roots in the scientific revolution and rationalist philosophy of the sixteenth and seventeenth centuries. Enlightenment figures Locke and Montesquieu, whose ideas figure prominently in the US written constitution, made no secret of their admiration of Newton and Descartes, who in turn became icons of liberal democratic thought. Science and rationalism provided a dual justification for liberal democracy—a political theory that would bring together empirically discovered knowledge of the external world with knowledge of self gained through Cartesian journeys of self-discovery. Liberal society would be a model of Newton's harmonious universe governed by natural laws divine in origin but discoverable through science, a society populated by rational individuals whose sure knowledge of their own interests could provide an efficacious foundation for collective political decision.

It is important to the constitution of liberal societies that science provides the one source of authority not defined in terms of offices or elective position—it represents "an" authority. But the ability of science, as an anomalous form of liberal authority, to be a practicing "in" authority for ongoing questions of politics and policy nevertheless is rarely certain, and its effectiveness should not, therefore, be overstated. Science wears the authority of revealed truth in liberal societies, it is true, but it is not a cloak necessarily draped over any specific individual members of the scientific community. More than this, within the community of science this denial of individual authority is a matter of catechismal doctrine, because, as R. S. Peters notes, it is never justifiable "in scientific institutions to set up individuals or bodies who will either be the originators of pronouncements or who will decide finally on the truth of pronouncements made."[39]

Scientific authority, then, is basic to liberal politics as its sole epistemocratic form, but it is often at odds with other sources of political ("in") authority. In such conflicts scientific authority is usually the loser; thus, trust in the authority of science among citizens should not be overstated. In the United States, for example, the institutionalized relationship of science and government has a short history, whose inception usually is dated with the publication of Vannevar Bush's report after World War II. Bush's report stimulated burgeoning government support of scientific research and the institutionalization of science into major government agencies. Yet science agencies such as the National Science Foundation (NSF) and the National Institutes of Health (NIH) have never held cabinet rank and are often losers in budget battles with other departments in the executive branch. Similarly, although for several decades Congress and various presidents rallied the nation around the Atoms for Peace program in the fifties, the space race in the sixties, and the war on cancer in the seventies, to name but a few examples, since the 1980s support for scientific research has fallen victim to budget cuts and lessened public support.[40]

Especially among liberal citizens in the late twentieth century, science no longer shines with the same allure as either the source of truth or the locus of epistemocratic political authority. As science and technology produced or educated about new risks, citizens have become less receptive to the use of science to achieve political goals or resolve political disputes. Although the urge to await technological fixes is still strong, the apparent impotence of science in delivering on its promises to eradicate hunger, poverty, crime, not to mention to provide "energy too cheap to meter," have tarnished science's claim to authority.

More important, however, the emptiness at the center of the liberal approach to authority, so clear to Arendt and contemporary conservative thinkers, has never been adequately filled by science. Even though among the three elements of liberal authority science alone as an epistemocratic variant offers the assurances of truth, it is a truth too often considered irrelevant to the issues of the day and to the citizens concerned with them. Nowhere is this clearer than in political disputes involving science and technology issues such as energy, waste disposal, medical technology (including reproductive technologies), and genetic research. The record of these controversies, carried out at every level of government and politics, is now familiar. At public hearings or informational meetings scientific experts give testimony; it is subsequently contradicted by other, equally credentialed professionals, and in the end scientific laypersons, either elected, appointed, or casting ballots, make the decision following their own, often scientifically ignorant, instincts. Price offers an explanation:

> As for science, it is in its abstract basis a reductionist approach to knowledge, with no concern for values. As a result, it is true that a strictly scientific analysis of a current policy issue always leaves major questions to be answered by moral or political judgment. You can measure the contribution to the economy to be made by strip-mining new veins of coal more precisely than you can measure the damage to the environment, including its aesthetic costs. You can measure or estimate the lives that could be saved by requiring the use of seat belts in autos, but the costs in terms of personal freedom cannot be expressed statistically.[41]

What has always been lacking in the liberal approach to authority is precisely what is most needed in the type of disputes to which Price refers. Most often in policy disputes the issue is redefined not as a scientific problem requiring technical expertise but as a moral issue turning on definitions of equity, morality, liberty, or rights.[42] Who are the moral authorities to inform the parties in dispute? To be blunt, there are none; liberalism does not grant authority over moral issues except to individual citizens, who themselves cannot agree. In the words of the California legislature after hearing 120 scientists argue over issues of nuclear safety, "the questions involved require value judgments, and the voter is no less equipped to make such judgments than the most brilliant Nobel Laureate."[43] No less equipped, perhaps, but no more successful either, at least in collectively setting policy for a nation.

Since Machiavelli changed the source of authority from knowledge of the correct ends of decision to the process by which decisions are made, liberalism has struggled to provide authoritative leadership concerning issues involving questions of public morality, questions that liberal citizens take just as seriously as the ancient Greeks. But whereas the Greeks stood ready to defer to demonstrated authority in moral issues, liberal citizens bridle at the loss of liberty to choose their own way that such deference presumes. To protect the autonomy of citizens, liberalism disperses and "chastens" political authority. In the process—and the process is the key—any sense of certainty is intentionally avoided. Science can be called upon for advice, but of course the appeal of its authority is largely based on the modesty of its claim to provide only theory and falsifiable knowledge. Yet many of the issues on the political agenda of liberal states today come from science itself and its technological developments. The risks of our technological age cry out for technical solutions that founder on moral issues; the politics of distributing technological risk offers legions of authoritative technical experts but no singularly authoritative moral voice. As Mark Warren says, "[D]ifferential risks and burdens are devastating to democratic authority, as the current cynicism about American democracy attests."[44]

Authority in Risk Politics

Even granting the epistemocratic elements supplied by its alliance with science, the version of political authority bequeathed by liberal theory is utterly unsuited to the needs of our technologically risky age. An authority with no center or focus, made tenuous by the temporal expectations of campaign politics and unable to speak to the moral content of contemporary political issues, cannot resolve the disputes generated by the demands of technological risk. Effective risk decisions require collective and normative judgments because they must mirror risk's emergent nature. But the redefinition of technical issues as moral ones that occurs within NIMBY manifestations has repeatedly denied the possibility of collective policy choice. Because the ontological foundation of liberalism enthrones each individual as an autonomous moral agent, moral issues become matters of intensely private interest that do not allow sacrifice to collective definitions of equity or justice. Without an external communal authority to whom an appeal can be lodged, we are back in the state of nature, all together and alone.

Within this authority vacuum a curious development is emerging, one that both witnesses to the absence of authority and to citizens' yearning for an authoritative voice to be heard. It is in one sense an epistemocratic "deferral" of authority yet at the same time manifests the populist refusal to acknowledge any one else as personally superior, a refusal upon which liberal egalitarianism is based and which, as de Tocqueville recognized, makes only a strictly "in" authority possible in liberal societies. Unlike the first deferral or entrusting of authority to scientific experts, this second "*re*personalizing" of authority is an investment in one's neighbors, fellow citizens directly involved with the production of risk. The politics of nuclear power supplies our example.

The story begins with "the ebbing of trust" in both nuclear companies and their government regulators. In Connecticut, as elsewhere, nuclear utilities have come under fire for widespread safety violations, many of which were either ignored or missed by regulators from the DOE or NRC. The severity of the breakdown in Connecticut led to a complete shutdown of all reactors in the state for extensive repairs by July 1996. Investigative stories in *Time* and *Newsweek* magazines in 1996 recounted a sordid history covering two decades of malfeasance and inattention to the safety and well-being of citizens, whose welfare both the industry and its regulators were supposed to ensure. The simple facts of nuclear technology are that it is a highly hazardous technology whose risks can nevertheless be managed safely, providing that safety procedures are devised, enforced, and precisely followed. Those procedures (with the exception of waste disposal) have largely been in place for decades; thus, questions of risk in essence boil down to the trustworthiness of industry executives and government regulators to follow procedures. If they do, risks are negligible; if they do not, nuclear technology as a whole generates levels of risk no rational observer should be expected to accept.[45] In the face of such obvious malfeasance on the part of executives and regulators, one would expect citizens to lose confidence and withdraw trust from these authorities.

The surprising reaction from those living closest to targeted nuclear installations is not that they withdrew trust from those with authority over nuclear safety matters. This did in fact occur as expected and extended both to those "in" authority—officials in the NRC and DOE—and to those "an" (or epistemocratic) authorities that provided expertise. But, unexpectedly, citizens expressed confidence and trust in their neighbors who actually worked at nuclear installations to protect them from actual harm. Here is trust detached from authority per se and attached to the perceived communal commitment of those who participate in a work-related way to the generation of risk. As the *New York Times* examined the phenomenon, its writer pointed out that

> increasingly, the confidence expressed is personal, based on faith in the friends who work on Millstone Point, rather than in Northeast Utilities or its watchdog, the Nuclear Regulatory Commission.
>
> "I have a philosophy about this that I think many people share," said David Cini, the First Selectman of East Lyme. "I know people at Millstone, and I believe that the hundreds of people who work there all want to be safe. These are not people who would allow things to happen to jeopardize the safety of their families and themselves."
>
> Laura Fitch, the 80-year-old president of the Waterford Historical society, put her opinion more bluntly. "We live with these people who work there, and I assume they know what they're doing," she said.[46]

We should be careful not to generalize overly much from examples provided from only one area of technological risk, but, nevertheless, the changing trust/authority relationship in evidence in nuclear issues raises interesting possibilities for reconstituting authority in liberal societies. The Connecticut example in a sense resuscitates the relationship posited by Arendt, MacIntyre, and other classically minded authority theorists between authority and community. For them, authority begins with a sense of shared values within the community that all citizens are presumed to accept and from which a mutual trust is built. On that foundation is constructed an epistemocratic justification for authority based on special knowledge or experience that only a few members of the community are expected to possess. In the Connecticut case the same reliance on shared values and a specialized knowledge is evident in citizens' deference to their peers at the plant, but the latters' knowledge does not qualify them as better or, in political science terms, as "elites." Their authority comes only from their place within the community that grants a specialized knowledge.

As we have seen, liberal societies achieve such levels of community with difficulty, if at all, and thus build authority on an altogether different ontological foundation. If however, that trust can be itself resuscitated by confrontation with risk, a new possibility for authority within modern liberal societies is imaginable. It is one that incorporates an epistemocratic component but with a rather low threshold of expertise—simply "working at the plant" is sufficient warrant. Combined with an expected level of care that comes from being part of the community, such "attentive nonexperts" both within and outside the plant can be expected—*trusted*—to exercise their authority in a responsible fashion.[47]

I will return to this possibility for a new liberal basis of authority in the conclusion of this chapter, but before I do so, it should be clear that the prospects for an authority based on such trust among citizens and their leaders are lessened by several institutional roadblocks thrown up by liberalism itself, most specifically the judiciary. Since liberalism historically has difficulty accepting community as an overriding political or social goal, liberal theorists since Locke have constructed authority on a very different, noncommunal basis. Indeed, liberal authority, as it has come to be exercised most often in liberal societies when making decisions about the distribution of risk, is of a kind that explicitly repudiates communal ties. Here we are speaking specifically of the "in" authority of members of the judiciary as interpreters of civil law. Within liberal societies the judicial process that delivers authoritative verdicts is explicitly not communal but adversarial. The increasing reliance on judicial determinations for making risk policy in evidence in countries like the United States ensures that a communal basis for authority will be difficult to achieve.

An important policy development over the past three decades has been that citizens have responded to liberalism's absence of authority to resolve NIMBY claims by using litigation either to procure compensation for loss due to risk or to block by judicial decree risk policy decisions that impose costs on them, such as living in proximity to risky installations. The results have been both an explosion of litigation concerning risk and a new policy reality that places courts in the authoritative position of making policy for risk. George L. Priest explains:

> Since the late 1960s, our civil justice system has adopted the premise that civil damage awards enforcing liability rules and statutory rights can optimally regulate *every* source of societal risk. Building from this premise, our civil courts have become the most powerful institution of the modern state for regulating risk.[48] (emphasis Priest's)

Since issues of justice or rights are central to many risk debates, it may seem logical that the legal system would become involved in adjudicating them, especially as something of a last resort given the vacuum of authoritative action elsewhere. However, relying on courts to make policy for technological risk is a bad idea for several reasons that go to the heart of the judicial system and to the nature of modern risk. Ultimately, those reasons also elucidate why political authority sufficient to resolve the many-faceted debates over the distribution of technological risk is lacking, and likely will never be supplied without a thorough overhaul of liberal politics.

As a general proposition, policy making by unelected public figures seems a patently undemocratic approach. This first reason is a common complaint against judicial activism, of course, but for decisions concerning the distribution of risk it carries special power. Because the risks at issue in many civil cases are without precedent in either the history of science or of law, judges have little guidance in how to apply rules of precedent or common law. This is a pronounced difference between several types of risk cases and almost any other legal action. For instance, siting disputes assume a different level of urgency when the risk posed by construction will persist not only for a generation or two (as in cases, say, of low-income housing) but for thousands of years (as in the case of nuclear waste disposal). Similarly, suits involving paternity or child custody have been forever changed when the definitions of maternity or paternity require prefatory adjectives like "genetic," "birth," "biological," or "donor." Without the guidance offered by prior decisions that can realistically be used even as analogues—even if not precisely as legal precedents—judges would need to exercise a degree of discretion over the future that far exceeds any extant definition of judicial activism.

Beyond this impact upon the common law heritage posed by new technological risks and the disputes that accompany them, other aspects of the day-to-day functioning of courts disqualify them as adequate authorities in disputes over technological risk. The policy-making power of any court waits upon a case being brought to it, of course, and the nature of any individual case may prejudice the policy effectiveness that emerges from its decision. This can happen in a number of ways. First, litigants in civil cases as a whole do not often manifest the same economic, ethnic, or educational characteristics as society as a whole. Law suits cost money to pursue and, despite the growing use of advertising by attorneys, rely on the public's awareness of what claims are actionable. Not surprisingly, citizens with more money and education are far more likely to initiate suits protecting their rights in the face of risk than citizens with less of either. It is no wonder then that technologically risky installations are more often sited in impoverished areas than affluent ones.

A second form of prejudice emerging from judicially based risk policy derives from what Peter W. Huber, quoting Nobel chemist Irving Langmuir, calls "pathological science: the science of things that aren't so."[49] Pathological science enters the courtroom in the form of alleged scientific expertise offered by credentialed scientific cranks (or "outlaws"), usually paid by attorneys representing either side for their "expert" testimony, which is either debatable or, simply, false. How can such witnesses be believed? First of all, Huber explains, many of the "most acrimonious and troublesome areas of risk litigation today lie at what courts have labeled the 'frontiers of knowledge.' Places where 'the factual finger points' but 'does not conclude.'"[50] Examples include the alleged effects of Agent Orange, the dangers of exposure to low-level electromagnetic fields, asbestos, radon, microwaves, and the alphabet soup of chemicals like PCBs, EDB, or the like. Scientists disagree widely on the degree of risk posed by such exposure, and thus expert testimony—especially if bought and paid for—can also vary to the point of contradiction. Yet courts "have become steadily more willing to decide factual issues that mainstream scientists consider unresolved, and to award judgments on the strength of scientific claims that few scientists would endorse."[51]

Indeed, today scientific expertise is so often before the bench that it remains a mystery why it so rarely is in evidence behind it. Few judges have extensive training in science, yet they are called upon to pass judgment and award huge monetary claims on the basis of very arcane scientific testimony. These cases can produce what Huber names "junk-science verdicts." The growing prevalence of such verdicts does not bode well for the continued authority of the courts in scientifically defined disputes, since, Huber says that "in the end the public acceptance of judicial power depends in considerable part on the shared belief that the courts tend to the facts right. When the real scientists catch up with the junk-science verdicts, as they inevitably will, courts will end up in disrepute."[52]

Slowly, judges are becoming aware of their deficiencies in resolving cases that turn on technical issues of fact versus "junk." As Justice Rebecca Kourlis of the Supreme Court of Colorado notes, reeducation of judges in the sciences is crucial today, for every case that comes before a judge involving scientific questions potentially sets a precedent for future decisions. Judges today are nervous about their verdicts in such cases, she admits, because "you always worry that you'll be blindsided, not see the novelty or complexity of an issue. You might treat it as something you've seen before and set up a bad precedent that could affect parties for a long time."[53] Lurking in the back of the minds of judges concerned about their lack of scientific expertise is the infamous judgment of Supreme Court Justice Oliver Wendell Holmes in upholding the enforced sterilization of retarded people on grounds that "three generations of imbeciles is enough."

Individual judges at present certainly have a knowledge problem in deciding risk cases—that is, they do not have enough; still, some reeducation efforts for judges are being initiated to remedy this lack of adequate preparation.[54] More troubling is the realization that liberal society as a whole has an epistemological problem in taking risk decisions to court in the first place. Judges cannot make good risk decisions because those decisions cannot be made in the ad hoc, adversarial environment of civil law cases. These cases presume a too individualistic, even solipsistic, ontology of risk: they postulate that litigants are discrete individuals with utterly separable and adversarial relationships relating to modern risk. But the emergent nature of modern risk calls us away from these assumptions. Technological risks alert us to the inseparability of our lives. These risks are the products of our lives together and of the collective decisions we make and have made about such things as energy, transportation, the distribution of scientific research money, or which threats to target for eradication. Our decisions have, in some instances, emerged over the course of a hundred years or more of development—development both of science and technology and of the lifestyles evident in society as a whole that rely upon them. They cannot be overturned in a single courtroom by an scientifically ill-informed judge in response to a claim pressed by a single individual against another.

Finally, the courtroom cannot be the source of authority over risk decisions because ultimately courtroom battles diminish the role of moral standards such as culpability or personal responsibility that such cases ostensibly rest on. This is an ironic consequence of the increased reliance on the courts for risk decisions, because individuals go to court explicitly to enlist the judge's acceptance of their moral position, based on their own interpretations of their rights. The growth in court time spent in litigation of risk issues has been the result of two new legal definitions. First is that "the predominant function of modern law is to allocate risk."[55] This is an inevitable consequence of the recognition of the role of risk and risk awareness in modern life and in a sense is not really substantively different from other definitions of social justice that stand at the foundation of any legal system. All notions of social justice have at their heart a conception of the fair allocation of something. It is merely (sadly?) a commentary on either the pessimism or the scientific sophistication of our time that for us that something is negative.

Second is the new definition within contemporary civil law of the concept of liability for risk. Legal liability is no longer associated with the issue of who causes risk but instead with who contributes to it. This focus on *contribution* rather than *causation* is a crucial reason for the explosion in risk litigation, because, by the definition of social life, every action taken by an individual potentially contributes to the risk faced by another, as George L. Priest explains:

> Our modern law, devoted to risk control, focuses less upon strict causation than upon contribution to the occurrence of the harm. . . . The new regime of risk control thus vastly expands the opportunity for the attachment of legal liability as well as the importance of civil law as an instrument of social control. . . . Our modern civil law encourages litigation as an instrument for internalizing costs to control risks. . . . Thus virtually every action by every citizen becomes subject to potential legal review because every action will increase the risk of some loss in some way.[56]

Defining liability in terms of contribution to risk indicts us all and thus makes issues of guilt or innocence in the exposure of others to risk largely irrelevant. We are all guilty, or at least no longer innocent. All that remains to be decided (in court) is who has to pay; issues of "the moral quality of one's personal actions" or of how to distinguish it from the quality of others' actions simply have no bearing—there is no authority left to whom to appeal for judgment.[57]

The emergent nature of technological risk confirms that indeed we are all complicit or, as we will see in chapter five, responsible, for the riskiness of modern life. But degrees of responsibility cannot be determined by courts through an adversarial process of award and penalty. What the courts have been determining—who gets and who pays the money—is largely irrelevant to the larger issue of how to make policy for risk distribution. That issue requires an authority the courts do not possess.

Conclusion: the Authority of Ends

The reversion to the courts as a last-ditch effort to arrive at risk decisions is proof of the failure of the liberal model of authority. Not only is this legalistic understanding of authority not up to the epistemological and intellectual demands of risk decision making, it also lamely proffers a solipsistic approach to a profoundly collective problem. Thus, it truly represents, as Hannah Arendt decreed, the demise of authority, the loss of a common world. In its place is a political reality populated by citizens primarily related to each other only legally rather than morally, since the shared moral feelings and definitions necessary for a moral community can be supplied only by "an" authority. But, as we have seen, liberalism's traditional candidate for "an" authority status—modern science—remains inadequate to the task.

There is considerable irony in this conclusion; indeed, citizens go to court in risk cases to redefine the issues as moral rather than scientific or technical. And they are correct in their intuition that the issues of risk are indeed moral in nature—today technological risks pose threats to our moral community in much the same way that low-tech murder or horse stealing did a hundred years ago. In Douglas MacLean's words,

> It is far from clear, however, that acting to reduce murder and assault is more effective in saving lives today than cleaning up environmental degradation. Today, because of changes in knowledge and ability resulting largely from new technologies, the deaths we decide to allow from pollution or from hazardous technologies or from occupational risks take on a new moral significance, for these decisions can now threaten the climate of life that is central to this conception of morality and community.[58]

But what is this "conception of morality and community" newly threatened by the risks of our technological age? It never was the moral community that Arendt envisioned as presided over by her preferred model of authority. It is the liberal version of community built upon interests and rights and presuming an adversarial relationship between citizens safe in their individual private realms. It is a community of individuals whose self-definition is as Kantian "ends in themselves." Whatever external authority they indulge is never allowed to change or threaten that status. Government is allowed to exercise authority over means—it can preside over the processes of liberal politics—but the real authority is that embodied by these "ends," these private individuals, us.

When these individuals take their concerns to court, it is with the view that their authority as ends in themselves is being insufficiently recognized by others. That is, they claim their rights are being violated, their dignity or "endship" is not being sufficiently honored. This is indeed a moral claim and requires a moral judgment, but, as we will explore in the next chapter, pressing a right is not the same as pursuing Arendt's (and Plato's) communal good. Liberalism has redefined morality as the realm of right rather than of good. Yet even this (for Arendt) lowered view of morality is threatened within the context of science and technology and the risks they pose. Such a populist "authority of ends" is, in this first sense, the authority claimed by Robert Paul Wolff's "anarchic" Kantian individuals who can never surrender the sense of their autonomy in the name of communal responsibility. This view of authority is in evidence wherever the NIMBY personality expresses itself, and it is the logical conclusion of what authority has become within both liberal society and liberal political theory.

A second, less conventionally liberal meaning of the authority of ends concerns modern technology itself. The risks of high technology are also about ends, but in a different sense—they portend possible outcomes that threaten citizens individually and collectively. It is these ends in the sense of consequences that are the focus of risk disputes, whether in courtrooms, in legislatures, or in the streets. As such, they possess an authority all their own. It is an authority that citizens (viewing themselves as ends in themselves) who stand to suffer the consequences of technological risks claim is their warrant to make the decision about siting regardless of the advice of those with expertise.[59] If some citizens (in the example of siting disputes) alone are in line to bear the authoritative results of such risks, they claim the right before all other citizens or officials to make decisions regarding them. If, in other words, the potential victims of risk have not the authority to determine risk's course, then the potential consequences of risk carry not merely authority, but tyranny.

It may seem odd to speak of an outcome or end as possessing authority, rather than an individual or office. Especially if we take the ends of authority in its first, Kantian sense to mean individuals as ends in themselves, viewing the consequences or ends of risk as also authority-bearing seems too consequentialist to fit within the same moral theory. So, at least, has always been the nature of the dispute between utilitarians and Kantian deontologists. Even though both conceptions of authority are recognizably liberal in their paternity, now it seems that liberalism has too many sources of authority rather than too few. The impact of modern risk on liberal thinking has many faces.

But notwithstanding such potentially interesting philosophical effects of risk, this second conception of the authority of ends is a version of authority that recommends itself in collective choice situations such as the making of risk policy. Because liberal citizens insist that authority must be grounded in their consent, modern technologies offer a choice of outcomes arising from their adoption or rejection. For instance, energy policy presents several different options to citizens, each of which carries its own costs and risks. The choice between technologies and risks may be made, according to Scheppele and Soltan, "not because there is a particular sort of leadership favoring it, but rather because of the attractiveness of the outcome independent of the source of the suggestion."[60] Thus, they conclude, the outcome itself can be seen as "authoritative."

What Scheppele and Soltan name the "authority of alternatives" is close to what I mean by the second meaning of the authority of ends, with three important differences. First, "alternatives" carries the presumption that all outcomes are equally possible and choosable. The emergent nature of risks, on the other hand, confirms that some ends will be inevitable no matter what we decide now, either because they have always been out of our control or because earlier, irreversible choices have closed off future ones. Nuclear power and long-term risks that come from radwaste are an obvious example.

Second, seeing the authority of ends as decisive in democratic policy making for risk will require that citizens themselves are empowered by that authority. Since NIMBY cases make it clear that citizens already have a sense of why the authority of consequences matters, part of the authority of ends must presume, in Warren's terminology, "attentive nonexperts." That is, citizens claiming the authority to be treated as ends because the ends or consequences of a risky technology are themselves potentially authoritative must be better informed themselves about the technologies and consequences they entail. They must assume added responsibility—as must society as a whole—to educate themselves about the risks over which they are claiming decision-making authority. Doing so will include both reeducation efforts as well as some deference to epistemocratic authority.

At this point the example of Connecticut residents trustingly deferring to their neighbors who work in nuclear installations is interesting to reconsider. The authority of those workers in the eyes of their neighbors is somewhat epistemocratic, but it also relies heavily on a communal element that validates those citizens' judgments at work as likely to have positive consequences for both them and their neighbors. The authority of the ends chosen at work has legitimacy precisely because those ends are chosen by individuals trusted by their neighbors to share their values in a communal fashion. Such trust (and therefore authority), in other words, is viewed as legitimate, partly due to the workers' expertise and partly due to their sharing of a "common world" with their neighbors. The authority of ends presumes, then, not only a normative authority, as Scheppele and Soltan also see as part of the authority of alternatives. Rather, the third difference between them is that the authority of ends is also explicitly a communalistic grounding of authority, based on more than the rationalistically perceived "attractive properties the alternative possesses."[61] An additional aspect of the authority of ends appeal is the extent to which it invokes and builds upon the trust and connectedness that comes only with a sense of community.

This twofold understanding of an authority of ends is a marked departure from and a reconceptualization of the old liberal "calculus of consent" to authority, and it holds some promise as a way of coping with the risks of technology. Yet we should avoid excess optimism, for this view of authority requires much of liberalism and from liberal citizens, especially in terms of a fuller commitment to the idea of community. To say that the ends or consequences of risk possess authority is to recognize not only their inevitability but also the difficulty of preventing their occurrence within the liberal framework of conflict resolution and policy making. Because these risks are emergent from the collective life of a community developing technologically, the negative consequences they threaten obviously cannot be avoided simply by following the authoritative decisions of courts making policy in an ad hoc, adversarial manner. Partly this is a matter of the failings of the judicial system recounted in the previous section. But the negative ends of technological risk are also partly unavoidable because the language of rights used in risk disputes fails to take into account the emergent or collective character of risks. Ultimately we must ask to what the authority of citizens as ends in themselves entitles them. If their rights render to each of them "an" authority, are they then entitled, as many civil actions have in fact accomplished, to stop socially valuable technological projects that pose risk to them? Or, as "an" authority are they merely entitled as a matter of right to, as Richard E. Flathman claims, the "distinctly respectful attention"[62] of others when they are in court?

What rights entitle citizens to accomplish in court or elsewhere in the face of technological risks is the topic of the next two chapters. Before moving on, however, three final points should be noted about the authority of modern risks. First, saying that risks have authority should not be taken as equivalent to the old "autonomous technology" thesis. Risks pose consequences (or ends) that cannot be evaded because they are the products of the collective actions of both the scientific community acting as a whole and of any particular society acting as a policy-making entity. Risks carry the authority emerging from the collective economic and development decisions of a society. They arose as science and technology advanced—an advance that was itself the product of countless individual *and* collective decisions concerning investment, education, and effort. Thus, risks ironically manifest, in a way not intended by Arendt or MacIntyre or anticipated by Friedman, the "shared beliefs" upon which "an" authority is constructed. Technological risks do not just happen; they are the products of human action. But, crucially, they are the product of *collective* human action—they are emergent phenomena. As such, risks are authoritative but never autonomous; they only *seem* autonomous to observers overly steeped in individualist ontology and epistemology—that is, to most of us within the liberal tradition.

Second, the authority of ends that risks possess stipulates that risks and their consequences are moral in nature. It may turn out to be the case that the moral nature of risks cannot be contained within a theory of rights, but that will only prove the insufficiency of rights by themselves as a moral language. We would suspect this already because we have seen how difficult it is to make risk policy in court. But the citizens who take others to court are on the right track in recognizing that the authority they need must be an authority able to determine the morality of risk's consequences. It must be, in short, a moral authority.

Finally, on the surface at least, it would appear that the moral authority of ends implies that a consequentialist moral theory is requisite for risk policy. If so, this implication will be another argument against the current liberal obsession with rights as the beginning and end of both morality and politics. Yet we should not let go of rights too quickly; even Mill understood that in some instances guarantees of safety should be seen as matters of right. Technological risks offer many instances of danger; what we have yet to determine is where and whether responding to the dangers can legitimate a denial of rights. All that is certain at this point is that traditional liberal versions of authority cannot answer this question.

Risk, Self-Interest, and the New Technology of Rights

By giving rise to the authority of ends, emergent risks and the politics they spawn have altered both the political and moral content of liberal theory. Politically, citizens who see themselves as ends in themselves, and who are confronting the disquieting consequences or risky ends of technology's implementation, tend, as we have seen, to seek the legalistic assurances that victory in the adversarial process of the court system can provide. In the heart of that process beats the idea of rights, and upon it is built the conception of individual autonomy that within risk politics challenges liberal authority. Risk policy as a result is most often made—if made at all—in court on the bases of judgments concerning the balancing of individual and collective rights.

Since technological risk plays a role in most aspects of politics today, this change of venue to the courts is a major consequence of the impact of modern risk on liberal politics. It is also one that, from a democratic point of view, should not be especially welcomed. As Cass Sunstein remarks, litigation is not the same as democratic politics: "Reliance on the courts diverts political energies and resources from democratic channels. The substitution of litigation for politics has large costs. Citizen mobilization is a public good, inculcating political commitments, broader understanding, and the practice of citizenship."[1] It is obvious that within the liberal courtroom the currency of such individualistic politics is the idea of rights. The disconcerting question that Sunnstein's remark poses is whether there is anything left beyond the idea of rights within the liberal view of politics that might resuscitate the practice of citizenship beyond the confines of the courtroom.

Answering that question invokes a deeper issue for liberal theory: what is there beyond or in addition to rights within the moral foundation of liberal society? Do rights constitute all of liberal moral theory, and, if they do, what hope is there for deciding the moral issues that arise around the consequences of modern technological risks? Those issues have for the most part resulted in political and policy stalemate, which puts not only society's material future at risk but also the rights of those who might in the future populate it. All of these issues constitute the broad concern of this chapter: what impact do modern risks have on the concept of rights and, as a consequence, on liberal ethics and politics?

In its first 10 amendments, the US Constitution guarantees to every citizen the signature rights of a liberal society—the rights to choose and take risks. The Bill of Rights also promises to protect each person from the resulting risks when others avail themselves of the same opportunities. Both guarantees extend only so far, of course. At stake are personal liberty on the one hand and the requisites of social life on the other. Here, as in many places in that remarkable document, liberty and necessity are conjoined in a muscular yet delicate balance. Rights always presume this equilibrium of risk and responsibility. Too little of the former robs rights of their value; an insufficiency of the latter abridges their effectiveness.

It may seem odd today to speak of rights as positively associated with risk taking. Liberty seems to reside more in the protection from risks than in our openness to them, and indeed in our technological age liberty seems much jeopardized by the risks of everyday life. But, of course, personal freedom is always a risky proposition, because a free person must rely on himself or herself if he or she is to insist on being unrestrained by others. As Sir Isaiah Berlin famously instructed, liberty has both a positive and negative side, and it is the positive sense of liberty that is often the most rewarding and the most dangerous. The same is true for risks, whether they result from the exercise of our right to speak freely, to believe what we will, or to associate with whom we please: we must have the right individually to assume these risks if we are to direct our own lives freely as autonomous beings.

Yet, as we have seen, the nature of modern technological risks makes such individualistic constructions of autonomy or liberty greatly suspect. Since modern risks emerge from the collective background of shared social life, the reductionist idea of an individual's right to assume a risk seems outdated; it seems to imply that accepting risk is a purely private decision without import for others. But if, as the previous chapters have shown, the risks themselves are not reducible to the individual action or people subsumed in them, how can individuals singly possess the right of choosing to accept or reject them? Can whole communities that require a joint solution to technological risks allow each citizen such a right and liberty to prevent collective choice? Then again, if it is unreasonable to so burden communities, does this then mean our constitutional rights simply cannot apply to the special kinds of risks we face today? And if that is so, then must we conclude that rights are a luxury no longer affordable in an era of emergent risk that permeates all of contemporary politics?

It would be an irony of historic proportions if the political rights spawned in the age of modern science were rendered moot by the technological fulfillment of that science's world-changing potential. Yet it cannot be denied that the risky products of modern science and technology (including simply more knowledge) pose unique and unanticipated challenges to the rights articulated in eighteenth-century documents such as the Declaration of Independence, the US Constitution, or the Declaration of the Rights of Man. This chapter first explores in detail how modern science and technology have altered discussions of even the most basic rights of life, privacy, speech, and association.

At the same time, modern life poses so many daunting risks that many citizens are tempted to wrap themselves up even more tightly in the security blanket of their allegedly undeniable rights. Doing so, of course, is the stuff of siting disputes. Indeed, the cry of NIMBY is rarely a plaintive call, but usually a full-voiced demand that my rights concerning my backyard—or my property, my children's' genetic endowment, my future—be respected. Rights make risk decisions difficult, and therein lies an irony of *futuristic* proportions. As we have seen, because modern risks are emergent in nature, they appear especially threatening in their sheer incomprehensibility to citizens schooled in the individualistic language of liberal politics. Thus, it is no surprise that citizens fall back on the absolutist language of rights to protect their interests. But the reductionist language and uncompromising demands of "rights talk"[2]—usually uttered, as discussed in the last chapter, in courtrooms—cannot define collective solutions for modern risks. In its very nature as emergent, then, modern risk propels liberal societies into disputes about rights, a context within which it is singularly difficult to find solutions for risk issues. This dialectic of contradiction and failure within the rights approach to risk policy is a second focus of this chapter.

Rights talk is under attack today from a variety of sources other than those concerned with making effective policy for technological risk. From within liberal theory utilitarians have never accepted the closure that the claiming of rights brings to discussions of public good. Contemporary failures of rights-based policy decisions further confirm to utilitarians the correctness of Bentham's admonition that rights talk is "terrorist language" threatening—and legitimating—violence to the common good.[3] Marxists of course still recall Marx's argument in "On the Jewish Question" that the whole idea of rights is an ideological blindfold hiding the oppressive nature of bourgeois society. Most pervasive still is the critique of rights that emanates from a variety of communitarian perspectives, beginning with Charles Taylor but including both liberal and conservative positions, and that also shares several features of the feminist theory argument about the gendered, isolating nature of rights images.

Given these new attacks upon the idea of rights, and conceding that, unlike authority, the concept of rights has been for centuries and still remains a fecund source of literature and philosophical argument, I need to offer a disclaimer for what follows. It would be impossible in one chapter to attempt to reinvent the concept of rights to fit what I contend is a politically changed reality wrought by technological risk. Similarly, it would be a foolhardy pretense that claimed to engage either all the literature or all the issues concerning the place and content of rights in contemporary life. So I disavow both attempt and pretense. What follows is a necessarily brief exploration of the ways in which modern risks have altered the historic debates about both what rights are and which rights it is legitimate for liberal citizens to claim. Within this exploration not all theories of rights can be addressed, but even a more cursory glance at rights theories than what I will provide leads to several conclusions, which are mine as well. First, risks have altered the criterion for who (or what) has rights. Second, risks have changed the meaning or content of several traditional rights, in terms of what those rights protect or protect against. Third, given the first two, making policy for risk requires a different foundation than that provided by the traditional "balancing" of rights among democratic citizens or between individuals and society.

Rights in Contemporary Theory and Politics

Since their annunciation in the seventeenth century, rights have been both politically and philosophically controversial items. Much of the philosophical dispute about the conceptual nature of rights need not concern us here, but we do need to explore certain aspects of rights theory relevant in the context of contemporary politics surrounding technological risk. Those aspects relate, first, to the realm or scope of rights—are they universal and whom (or what) do they cover? Second, what is the essential relationship that rights construct between individuals—do rights primarily erect claims, elevate certain interests, or impose duties? Third, what requirements do rights impose for rights-bearers—and who (or what) is eligible for inclusion into the category of those with rights?

Most contemporary rights theorists discuss all three of these aspects as central features of any rights theory, and it is the different degrees of emphasis on any or all of them that tends to distinguish different theories. These philosophical disputes are not our chief focus here, though I note them. Of more central concern is the fact that citizens often make claims about their rights in the face of technological change, and, as we will see, the impact of technology itself often affects what they construe to be the substance of the rights they hold. Although the substance of rights to, for instance, life or free association may be markedly altered by the advance of technology, we need first to explore how contemporary rights theory construes the conceptual basis of rights as defining a certain structural relationship between persons, society, and their environment.

To begin, it should be clear that rights in our usage here refer to *moral*, rather than legal or civil, rights. Legal and civil rights are those enunciated and protected in law by governments and polities and may, therefore, vary across different cultures. Legal rights are sometimes presumed to be the more general, moral rights, but, as L. W. Sumner notes, "moral rights are not simply legal rights, or conventional rights of any kind." We can easily envision legally accepted violations of what we consider to be unquestioned moral rights. Apartheid laws in South Africa and Jim Crow laws in the post–Civil War South are obvious examples. Moral rights encompass legal rights in Sumner's view, either providing "moral reasons for establishing or maintaining conventional rights" or for altering legal definitions to accommodate heretofore denied moral rights.[4] Moral rights, in other words, are taken in a universal sense to apply to all human beings and are characterized as imposing a correlative duty on others, regardless of whether the law of any particular society acknowledges either the right or obligation.[5]

Moral rights are the type invoked as self-evident truths in the Declaration of Independence and given substance in the Bill of Rights. In US political culture, as a result, these moral rights have become legal and civil rights as well; thus, the distinction between the two types is somewhat blurred. For Judith Jarvis Thomson, such blurring indicates the weakness of what she names the "Two Species" thesis in the first place. She insists that logically there can be only one kind of right, in that all rights must impose the same relationship between rights-bearers and rights-respecters.[6] What is different, she claims, is the *source* of rights. Legal rights clearly have their source in law; what is the source of moral rights? Interestingly, believers in the distinctiveness of moral rights as uniquely "universal" as compared to civil rights often disagree concerning what the source of those rights is, other than they are do not emanate from laws but from something above or "prior to" human-made law. For Sumner the source is a notion of universal utility; John Locke attributed moral rights as coming from God or nature; Jefferson in the Declaration contended they were both endowed by a "creator" and self-evident, thereby implying the source of rights as both divine and (perhaps divinely) logical; contemporary rights theorist Robert Nozick simply avoids the issue altogether. Nevertheless, in contemporary rights theory the distinction is usually maintained, even if, as in Thomson's case, disputed. I focus on the conventionally understood moral rights, since they presume the universal application of most proclamations about rights. Furthermore, as we will see, the effect of modern risk on the notion of rights is to challenge explicitly this universalistic presumption of moral rights.

Moral rights have gone under different names in the past three centuries. In the seventeenth century Locke and Rousseau referred to them as "natural rights" in order to call attention to their place in the modern notion of natural law. Calling them "natural" provided automatic justification for rights, both because they were an implicit part of human nature and also therefore of the "natural" order.

One hears little of natural rights today; instead, usage has moved to a somewhat different notion of "human" rights. Jeremy Waldron comments on the significance of the shift:

> At the level of rhetoric, the shift from "natural rights" to "human rights" marks a loss of faith in our ability to justify rights on the basis of truths about *human* nature. To call them *human* is now to characterize the scope of the claims being made rather than hint at anything about their justification. The term refers to universality and a commitment to equality and non-discrimination, particularly across races and peoples. But unlike "natural" it leaves open the question of justification or, worse still, takes the mere existence of a broad consensus on these matters to be a sufficient reason for avoiding the task of justification altogether.[7]

When justification of rights is at issue, some philosophers prefer the term "inalienable rights," as did Tom Paine and Jefferson. While universality and naturalness are also presumed in these constructions, inalienable rights focus on the holder of rights rather than on the respecter(s) of rights. As Diana Meyers uses the term, seeing rights as inalienable allows us to defend them logically in terms of their importance to moral systems generally, rather than on empirically arguable claims as to rights' universality or absoluteness. A "satisfactory moral system" is impossible without the recognition of inalienable rights, she argues, since a moral system would be self-defeating if it allowed the destruction of moral agents.[8] The inalienability aspect of rights is what protects individuals from such a fate—whether they are willing sacrificial subjects or not. Thus, to the extent that moral systems are desired and possible, inalienability must be the essential feature of moral rights.[9]

Perceiving rights as inalienable and therefore indispensable for moral systems calls attention to another important feature of rights generally agreed upon by rights theorists. As Joseph Raz argues, rights presume a context of reciprocity between rights holders and rights-respecters, such that the exercise of rights guarantees that others will accept a duty at least not to interfere. The "reciprocity thesis" highlights the essentially social nature of rights—rights can truly be said to exist only in the presence of their respecters. For Raz, this means that "only members of 'the same moral community' can have rights."[10]

What constitutes Raz's "same moral community" can, of course, be interpreted narrowly to mean only members "whose obligations to each other are thought to derive from a social contract"[11] or broadly to include all living human beings, as in the Declaration of Independence or the Charter of the United Nations. In either case assumedly rights are relevant only as social facts, because the existence of others is necessary for the acknowledgment of a right. But mere existence as a human does not automatically guarantee rights in Raz's more narrow construction of a moral community; he adds that a certain "capacity" for rights is also necessary for an individual to be said to possess them.

Many philosophical disputes over whether rights exist or are at issue in a given context depend on this concept of capacity for rights. As Sumner points out, this dispute erects two different models of rights: an agency (or choice) model that construes rights at bottom as a type of claim and an interest model that begins with rights as representing special kinds of interests that warrant protection whether they are claimed or not.[12] Raz tries to dissolve the distinction by arguing, first, that the person possesses certain "essential" interests and, second, that the possession of rights also presumes a capacity for them. Essential interests are those without which an individual's survival is jeopardized; thus, they have an "ultimate value" for the individual in that they are not derivative from any more basic interests. Second, capacity for rights presumes the ability to participate in a moral community as "promiser"—that is, one must be functionally competent to make claims and agreements and to participate as an active citizen. By Raz's definition some people—and all animals—do not have sufficient capacity; for example, "small children and some mentally deranged people lack it."[13] For Meyers, capacity is also defined by the needs of moral systems—the ability to press and respect claims of right—as well as by the ability to recognize essential interests. Rights are inalienable, she claims, if what they protect is an object that it is never obligatory to relinquish and for which doing so would in every case be an act of supererogation.[14] Thus, individuals having the capacity for rights are able to recognize their essential interests sufficiently well to make a decision about whether to claim as a matter of right the protection of an interest or altruistically to forego protection. For Meyers, then, the class of rights-bearers excludes animals and human fetuses.

Underlying the notion of a qualifying capacity for rights is the idea that the possession of rights presumes the ability to *do* something. Richard Flathman makes this plain in his focus on the "practice" of rights. The practice of rights presumes two things: first, that persons with rights take certain actions; second, that certain rules apply to the actions taken.[15] The ability to participate in such practices associates the notion of rights with that of individual liberty and, Flathman claims, with that of community as well, since claiming requires the use of common language and rules, as well as a joint acknowledgment of common authority.[16] The impact of rights-based arguments and moral systems on the spirit of community is much debated today as we will see, but at a minimal level at least, Flathman is correct that the exercise of rights presumes a communal agreement about language, rules, and practices.

Joel Feinberg further associates the *activity* of rights with another important value implicit in most arguments in defense of rights. Because the capacity for rights intrinsically requires the ability to press claims, Feinberg speculates that the very basis for human dignity and self-respect lies in this capacity. In fact, Feinberg goes so far as to suggest that respect for rights and respect for others are identical sentiments and practices:

> Having rights, of course, makes claiming possible; but it is claiming that gives rights their special moral significance. This feature of rights is connected in a way with the customary rhetoric about what it is to be a human being. Having rights enables us to "stand up like men," to look others in the eye, and to feel in some fundamental way the equal of anyone. To think of oneself as the holder of rights is not to be unduly but properly proud, to have that minimal self-respect that is necessary to be worthy of the love and esteem of others. Indeed, respect for persons (this is an intriguing idea) may simply be respect for their rights, so that there cannot be the one without the other; and what is called 'human dignity" may simply be the recognizable capacity to assert claims.[17]

This understanding of rights locating an individual in a set of practices or activities that includes other members of a definable community as well makes rights fully and intrinsically political things, more than merely social. This is especially true when having rights means that one has the capacity to assert claims. And, since liberal politics is defined in terms of conflicts of interest and the exercise of power, it is clear why rights are so central a concept within liberal theory. Rights as construed by Feinberg, Flathman, Raz, and most modern theorists presume that rights are necessary because certain essential interests of individual persons are in conflict. Furthermore, the claiming of such rights is a political act both for its public character and its relationship to power. A right is power over others' behavior—*legitimate* power made so by the presence of an essential interest deserving protection.[18]

If Feinberg is correct that it is the claiming activity that marks rights as particularly significant for the achievement of human dignity, it is clear why even liberals predisposed to reject rights, that is, utilitarians, have such difficulty doing so. All liberals accept the dignity of the individual as a paramount political value. If it turns out that dignity is rooted in the equal power evidenced in the public display of the claiming capacity, then liberal politics is difficult to picture without such displays. This may indeed be why even a utilitarian with the sterling credentials of J. S. Mill succeeds in ridding politics of natural rights only by substituting rights legitimated by rules (i.e., utile). Mill struggles mightily to defend individual liberty, as well as the equal dignity of women, without reverting to the language of natural rights, yet the pages of both *On Liberty* and *On the Subjection of Women* are filled with references to the rights that individuals possess as the products of social agreement.[19]

Together, the liberal focus on rights and its perception of politics as conflict give to Feinberg's image of "properly proud" citizens exercising their claiming power the look of adversaries engaged in (civilized) competition. Ronald Dworkin embraces this adversarial connotation for rights in his formulation of individual rights as "trumps."[20] Undeniably, Dworkin's image of rights is a powerful one within contemporary politics, especially the politics of technological risk. Success in blocking governmental or industrial plans in siting controversies often depends on the compelling force of the language of rights, and rights are meant in such situations as trump cards protecting the individual or the powerless in the face of superior power.

In the next two sections of this chapter I explore the effect of risk on the language and exercise of rights, as well as the effect of rights talk on the policy process for risk. Throughout these discussions the image of rights-as-trumps will powerfully lurk, but I want to suggest nevertheless a slightly different metaphor for rights that will become important for our discussion in chapter six of how to reformulate risk issues. If social processes are analogous to games of chance—as Dworkin's usage suggests—perhaps rights ought to be seen not as trumps but (to mix card metaphors somewhat) as "antes." That is, rights-as-antes denotes that, as in a poker game where one must "ante-up" before being allowed to play the hand, individuals have the wherewithal—the qualifying stake—always to enter the game of public discussion and policy making. They have the *right* to do so as equal members of society with essential interests at stake. Thus, individuals possess a guaranteed entry to the game, but, as their stake is not a trump, they cannot individually determine the outcome of any particular play. Because their rights are but antes, they have the right to play, to be heard, to participate, but not necessarily to win.

Indeed, as in a game of cards, there can be but one "winner" in every hand. The outcome of the entire game, on the other hand, can still be equal. Furthermore, viewing rights as antes means citizens' rights come into play at the beginning of every hand, not at the end; they make the game proceed according to acceptable rules but do not determine the outcome. Rights-as-antes guarantee a hearing for one's position and thus are of considerable importance to individual liberty as well as to the process of democracy, but they do not ensure that one's point of view will always carry the day.

The ante view of rights stresses even more than Raz's reciprocity thesis the social character of the concept of rights, as well as its presumption of essential interests and an ability to press claims as important features of rights generally. Thus, it shares important features of the contemporary theories of rights reviewed here. Yet it is fundamentally different from the briefly discussed other theories that share these same aspects of rights. The concept of rights-as-antes presumes that rights are "moral," but not universally so.[21] Whatever moral compunction rights exercise takes place only within a context or moral community that respects them. Rights-as-antes presume not universality but a specific democratic context in which citizens participate in what Simone Chambers calls "discourse ethics."

Following Habermas, Chambers roots all social rules or norms, including rights, within a practical discourse that provides them with meaning as well as whatever moral bindingness they possess. Doing so makes rights context-dependent instead of universally implied either by God, nature, or the logical necessities of moral systems. Additionally, rooting rights within the idea of discourse confirms their essentially democratic nature, because it makes citizens as a whole the final arbiters of what rights are and when they are in effect. Chambers says that

> [p]ractical discourse is a process of consensual will-formation, the object of which is a "generalizable interest. . . ." Within discourse ethics, practical discourse is the procedure undertaken to test whether a norm or general principle of action is rationally acceptable to all those affected by the norm. If a norm is acceptable under the special conditions of discourse then it is said to be valid. . . . Discourse ethics in its strongest version implies that a right is justified only when it rests on a rational consensus. Citizens themselves are understood as the final court of appeal in questions of rights.[22]

Within Habermas's "ideal speech situation"—for our purposes an ideal context for rights—the rights of citizens are those that make democratic discourse possible and guarantee that it continue. Thus, rights can never "end" a discussion about what is morally acceptable; they can only begin or continue that discussion and guarantee to all citizens the opportunity to be heard within it. If discourse is analogous to a game, rights are each player's entree into the play, their "ante." All players hold three such rights-as-antes: to freedom, equality, and mutual respect, as Chambers explains:

> The conditions of discourse include individual freedom, equality, and mutual respect: all individuals must have the freedoms necessary to participate in discourse; all must be treated as equal partners in the process of will-formation; no one may be coerced, pressured, manipulated or deceived into accepting an argument.[23]

In the remaining sections of this chapter, I discuss how technological developments and their attendant risks have altered both the substance and exercise of traditional liberal rights-as-trumps. So transformed has our world become that our rights—whether natural, human, or "inalienable," can no longer resemble their seventeenth-century forbears in content or in operation. Instead, our technologically risky world presents a context within which the decisions we must be able to make concerning risk seriously constrict the number of rights that we can inalienably claim. Rights that inhibit the making of those decisions are invalid. But, I argue, the rights that remain to us—our rights-as-antes—are essential to democracy.

Democracy and democratic practices are also essential to rights-as-antes. Such rights cannot guarantee for all citizens an equal outcome from any particular decision. Therefore, the egalitarian guarantees of respect and liberty promised by democracy are clearly necessary for protecting *all* individuals' essential interests at stake in risk decisions. Such a view of rights makes living with risk possible and, within a context of democratic community, morally responsible as well.

Technological Life, Liberty, and Pursuit of Happiness

When Jefferson named the rights of life, liberty, and the pursuit of happiness in the Declaration, he did not bother to offer definitions for any of these "self-evident" terms. He surely understood that liberty and happiness were subject to myriad meanings and understandings, but no doubt he felt everyone could agree at least on the meaning of the term "life." Most Americans today still believe that a right to life exists but are no longer in complete agreement about whether it belongs in the same sentence as other rights such as the right to "choose" or to "pursue" a certain quality of life. "Right to life" now indicates a particular ideological position incompatible with belief in the liberty to choose abortion or, for some of its holders, even with certain excessively carnal "pursuits." What has changed since 1776 that has so altered our appreciation of "self-evident truths" regarding rights, liberties, even life itself?

Contemporary understandings of the substantive content of basic rights are surely different today than when they were listed in the first eight amendments to the Constitution. Many social and political developments have contributed to the changed nature of rights, and, to be blunt, life in contemporary America would in many ways be unrecognizable to the constitutional framers. All that is truistic, of course, and I do not want to face a charge of reductionism, yet a major influence on the evolution of the meaning of basic rights has been the impact of modern technologies and the new opportunities, as well as risks, they pose for the exercise of constitutional rights. Robert E. McGinn goes further, arguing that modern technology has made an anachronism of the old, universalistic conception of moral rights; he also provides an evocative example.[24]

McGinn traces the developmental and legal history of the use of automated telephone dialers and describes how such devices can "impede or disrupt societal functioning" to an extent requiring abridgment of their use even in the face of charges that doing so violates the constitutional guarantee of freedom of speech. He notes that with 20,000 such machines in use during any given day in the US, it is possible daily to dial 20 million numbers around the country. McGinn reports that as a consequence "of the potency and number of autodiallers, significant communications breakdowns have already occurred."[25] In response to resultant constituent outrage, Congress in 1991 passed the Telephone Consumer Protection Act that banned the use of autodiallers for calling private homes. However, under litigation brought by the American Civil Liberties Union, a US District Court found the act in violation of free speech rights and blocked its enforcement.

McGinn's example supports his argument that the unbridled pursuit of technological development has diminishing returns for the quality of life in modern societies and requires that old universalistic conceptions of values such as rights be replaced by new, contextually defined and severely limited ones. He concludes his critique of "technical maximality" with the following discussion of its effect on the idea of rights:

> In sum, we need a contextualised theory of human rights. An acceptable theory of rights in contemporary technological society must be able to take on board the implications of their exercise in a context in which a rapidly changing, potent technological arsenal is diffused throughout a populous, materialistic, democratic society. Use of such a technological arsenal by a large and growing number of rights holders has considerable potential for diluting or diminishing societal quality of life. Indeed, insistence on untrammeled, entitled use of potent or pervasive technics by a large number of individuals can be self-defeating, e.g., by yielding a state of social affairs incompatible with other social goals whose realization the group also values.[26]

McGinn's argument is a powerful one, yet we should not rush to embrace its conclusion that the Bill of Rights is a charming but now irrelevant and unaffordable relic. Before we do so, I want to look further at four basic rights that Americans presume are their constitutional legacy, two of which are not actually mentioned in the Constitution and two that are: the right to life, to privacy, to freedom of speech, and to freedom of assembly (or association). I take these four by way of example only, since the impact of modern technology on other rights is also palpable. The use of modern computer technology in data gathering and its impact on the constitutional right to be free from unreasonable searches and seizures would be another example. Of course every age is capable of inventing its own technologies that violate the Fourth Amendment injunction against "cruel and unusual punishment." But more later about television.

The Right to Life

Unstated in the Constitution, the right to life is as much a "penumbral" doctrine surrounding the whole idea of individual rights as is the right to privacy under Justice William O. Douglas's famous formulation in *Griswold v. Connecticut*. Nevertheless, philosophers argue over the exact meaning and extent of this right, and politically the uncertainty of its power as a claim upon others extends beyond the abortion issue. As a right it has been uniquely affected by the emergence of modern technology and its attendant risks.

The right to life is generally taken to mean one of two things. First, as narrowly formulated by Meyers as an inalienable right, it means simply the right not to be killed. As a right it is defensible in her view by the truism that one cannot be a moral agent unless alive. For Feinberg, this obvious fact means that the right to life also is a valid claim "to be rescued whenever this can be done at reasonable cost."[27] Feinberg's addition broadens the right to life somewhat, although just how far is a matter of considerable dispute,[28] and in any case it does not depart in kind from Meyers's definition.

A second contemporary interpretation of the right to life is markedly different in that it adds some notion of *quality* of life into the equation of legitimate claims and concomitant obligations. Thus, it becomes, as Feinberg notes, a "right to live decently." Locke invoked this notion of the right to life in his insistence that it entitled one "to a life fit for the dignity of man." Hugo Bedau concurs that "the life to which we now think men are entitled as of right is not a right at the barest level sufficient to stave off an untimely death; rather it is a life sufficient for self-respect; relief from needless drudgery, and opportunity for the release of productive energy."[29]

Bedau's viewpoint is somewhat out of favor politically these days and in any case can be criticized as conflating several different rights and merging them illegitimately into one mega-right. Judith Jarvis Thomson concludes, for instance, that a definition of the right to life such as Bedau's really refers to a cluster of rights and that "a cluster-right is of course not a right in the strictest sense, since it is not correlative with a duty."[30] But under either the Meyers/Feinberg or the Bedau characterizations, the meaning of the right to life has been radically affected by twentieth-century scientific and technological advance.

Scientific understandings and technological abilities pose two different challenges to the idea of a right to life. Under the first definition of that right (as in Meyers's and Feinberg's), modern medical knowledge and techniques cast doubt on the very definition of life itself, and, critically, of death also. Secondly, if quality of life is put into the concept of the right to life, as in Bedau's characterization, then modern technology exercises enormous defining influence in its ability to deliver cheaply into every individual life what would have once been considered untold riches. Let us explore further these two conceptual impacts of modern technology.

Any consideration of the right to life proceeds under a presumption that one knows with an acceptable level of certainty when life is in fact present and when not. Current disputes over abortion and euthanasia show just how tenuous that knowledge can be. Without getting excessively drawn into either controversy, let us be clear that what is at issue in both cases is the right to live or die and therefore also what constitutes either the beginning or ending of life.

The abortion dispute did not begin with the advent of modern medical technology; abortions have long been a part of the reproductive history of the human species. But medical capabilities today in the area of reproductive technologies have already vastly changed human reproduction in developed societies and in the process brought revolutions in cultural and family life. Amniocentesis, cryopreservation of human embryos, ultrasound scanning, genetic screening, the "abortion pill," fetal surgery—these are all technologies affecting the reproduction of life. And, let's just say it, they all presume that abortions are a legal option.[31] Thus, they raise the issue of when life begins—of what life is. Abortion is indeed about the right to life under either of its two definitions and depending on whether the focus is on the fetus or the mother. Modern technology has put the right itself in jeopardy by pointing out just how difficult it is to define the right's object—life itself.

The same is true for the issue of euthanasia, the right to die. It is medically possible today to keep a heart beating in an otherwise "dead" body for over 200 days. The lungs can be made to function and respiration proceeds. Is such a respiring entity living? To complicate matters further, assume the presence of a growing fetus within the body plugged into the "life"-prolonging machinery. In several instances the bodies of pregnant women have been artificially kept "alive" long after brain death in order to provide as close to a natural birth environment as possible. Are such babies born motherless or not? When does life end, and do we individually have a right to end it? Living wills can now legally ensure the "right to die" in some states, but the moral—and political—muddle persists.

A primary reason for why it is so difficult to define "life" in cases involving abortion or euthanasia is that defining life so as to make legal and political solutions possible involves some redefinition of what *quality* of life is acceptable. In other words, modern medical technology often makes defining life in the first conceptualization of the right to life contingent upon providing a definition of life in Bedau's second sense of a "right to live decently." If we are repelled by the treatment of a women's body being used as an incubator after her brain has stopped functioning, it is because such life is not one "fit for the dignity" of the individual. Furthermore, reproductive technologies such as amniocentesis or genetic screening that presume the legality of abortion may, if made mandatory, indeed be able to deliver a species free of genetically transmitted debilitating diseases, but at the cost of reproductive liberty for those discovered to be at risk as carriers or victims of the diseases. Their quality of life then is affected negatively, if they wish to reproduce in spite of the risk. Conversely, if abortions are made unavailable, the economic costs in terms of guaranteeing a certain quality of life for vast numbers of children—and for mothers no longer able to choose—may prove prohibitive.

Similarly for euthanasia, it may increasingly be possible technologically to extend life, but it is extremely expensive economically to do so. Whether we are talking merely of extending to each individual the full benefit of modern medical technology or, more disconcertingly, of keeping the machines plugged in to extend life in some more controversial sense, the costs are enormous. Ultimately, of course, all citizens pay for the escalating medical costs incurred either by universal health care or by bodies attached to life-extending machinery. We all pay economically in insurance bills, medical bills, or in taxes. The moral price is more difficult to estimate.

If nothing else, modern medical technology teaches us that any definition of life must recognize the intimate interconnectedness of individual lives. This is a new lesson, and one that perhaps has little place for a notion of a right to life as traditionally conceived in terms of some kind of trump.[32] The interconnections may even form a "context" not amenable to the whole idea of rights as a basis for social choice. Before modern medical knowledge and technologies were available, these aspects of a so-called right to life simply did not exist because the definition of life was not an issue. In the presence of these technologies, then, new risks to that right arise, at least if one views it in a strictly individualistic, adversarial way.

The Right to Privacy

Prominent among the several reasons for why the right to privacy is never mentioned in the Constitution is that the founders considered the entire concept of "private citizen" a dangerous oxymoron. As liberals, Madison and Hamilton surely understood that an important function of government was the protection of individual liberty to pursue private ends. Yet the pursuit of self-interest could not remain completely within the private domain without resulting in factionalism, in their view the scourge of all democratic societies throughout history. The inevitability of factions meant the "public good is disregarded in the conflicts of rival parties," leading to effects "adverse to the rights of other citizens, or to the permanent and aggregate interests of the community."[33] Because, Madison argued, factions could never be avoided within a free society, the cure was to make private concerns public, to ensure that citizens would bring their private interests into the light of public debate. Thus, the pluralist theory that *Federalist #10* elaborates begins with the civic republican assumption that in the healthy operation of politics private ends are made public, and within this process of exposure and publicity the promise of liberty is realized and merged with the demands for justice and even virtue as the requirements of democratic citizenship.[34]

Similarly, Jefferson, though ever the defender of the private rights of the individual, admonishes that without citizens taking upon themselves the extremely onerous public responsibility to rebel periodically, the "tree of liberty" is without its "natural manure." Jefferson was no believer in the infallibility of the people when viewed individually as private, property-oriented consumers. But when seen as a whole—as a body of citizens, the public body politic—their combined wisdom and action are enough to protect the republican legacy of freedom. His famous commentary on Shay's Rebellion is a call for public engagement:

> God forbid we should ever be twenty years without such a rebellion. The people cannot be all, and always, well-informed. The part which is wrong will be discontented in proportion to the importance of the facts they misconceive. If they remain quiet under such misconceptions, it is a lethargy, the forerunner of death to the public liberty.[35]

It is true, of course, and later discussed in *Griswold* by William O. Douglas, that the rights defended in the Constitution imply a fundamental right to privacy as part of their implication that government should be limited in its power over individuals. But neither the founders nor several generations of political leaders and jurists felt it necessary to make this penumbral idea a separate, legal right until the intervention of modern technology in the late nineteenth century, an intervention compounded by inventions of our day.

As legal scholar Mary Ann Glendon records, the initial causes behind the juridical discovery of a legal right to privacy were the inventions of instantaneous photography and the means of mass communication. So great was the outcry among the rich and famous after seeing their pictures splashed across the pages of the daily tabloids that even the more respectable *New York Times* lamented the new "Ordeal of the camera."[36] The excesses of those forerunners of modern paparazzi led, Glendon believes, to the legal argument supporting a new right derived from the Lockean idea of property rights. Written in 1890 by attorneys Louis D. Brandeis and Samuel D. Warren, a *Harvard Law Review* article entitled "The Right to Privacy" argued that "[t]he principle which protects personal writings and any other productions of the intellect or of the emotions is the right to privacy, and the law has no new principle to formulate when it extends this protection to personal appearance, sayings, acts, and to personal relations, domestic or otherwise.[37] Brandeis would invoke the principle for the first time as a dissenting Supreme Court Justice 25 years later, arguing for a "right to be let alone—the most comprehensive of rights and the right most valued by civilized men."[38]

Today the origins of the legal right to privacy are most often traced to the Supreme Court cases considering the reproductive rights of individuals. First, in *Griswold v. Connecticut* the Court acknowledged a right to privacy residing in families, but through *Eisenstadt v. Baird* and culminating in *Roe V. Wade*, the Court finally acknowledged the right as one held by every individual regardless of marital status.[39]

It may be too much to conclude that reproductive or earlier technologies themselves directly led to the contemporary legal conception of the right to privacy. But it is undeniable that, when joined with other technologies capable of enlarging the capacity to intrude on the private lives of individuals, the right of privacy is the product of a technological age. Here, then, is a new technology of rights themselves, where technology defines what the right itself protects, that is, what the substance of the right is. Alongside the development of these new technologies capable of intrusion, the meaning of the right to privacy is still evolving in response to this growing technological capacity to eavesdrop, invade, and otherwise violate the privacy of citizens.

What then does the right to privacy mean in contemporary politics, and what is its effect on that politics? Thomson notes that "the most striking thing about the right to privacy is that nobody seems to have any clear idea what it is;"[40] yet this lapse seems to stop no one from claiming its abridgment. Thomson argues intriguingly—and correctly, I think—that the right to privacy is more subject than other rights to a particular mistake about what constitutes a violation of rights. Often cases that are claimed to be violations of rights are simply instances of "bad behavior," behavior that may be ungenerous or even invasive but not actually illegal or immoral in terms of rights violations.

Privacy rights arguments or claims are particularly prone to this misperception, Thomson argues, because if there is a right to privacy, it too exists as a "cluster" of rights, and "not a distinct cluster of rights but itself intersects with the cluster of rights which the right over the person consists in and also with the cluster of rights which owning property consists in."[41] Thus, the confusion over what actually is protected as privacy spills over into arguments over rights to life, property, speech, and others. In a sense, Thomson is only affirming what William O. Douglas said in *Griswold*: the right to privacy is a penumbral doctrine that embraces other rights; therefore, it is, as Brandeis noted, the "most comprehensive of rights." Yet recognizing these "clustered" aspects does not necessarily prepare one for the sheer number of instances in which the right to privacy is invoked in contemporary politics and law. Only living in America and experiencing firsthand, as de Tocqueville did over a hundred years ago, its "obsession with equality" can prepare one for the uses to which the right of privacy are put today.

It is difficult to avoid the conclusion that for most US citizens today the meaning of privacy rights lies in their legally protected opportunities to ignore the social effects of their private activities. Whether those activities have to do with sexual behavior and reproduction, communication, religion, recreation, the pursuit of economic interests, or a host of other activities, the right of privacy appears for most citizens to mean that whatever the social consequences of protected behavior, they as individuals are legally allowed to ignore them, and no one else—including government—can attempt to use the consequences as grounds upon which to prohibit the behavior.

In other words, the right of privacy allows individuals to avoid bringing their private interests into the political realm. It also denies to others the liberty to force one's own private concerns into public discourse—to "out" someone, for instance. Now, the right to privacy is certainly part of the bedrock of individual liberty, but, as Madison foresaw, it also can ultimately threaten liberty through the corrosive power of faction. Taken to extremes, the right to privacy allows public discourse to atrophy as private interests are increasingly pursued outside of the public eye and citizens refuse to participate in collective decisions. Because of modern technologies ranging from instant cameras to more sophisticated surveillance and control devices, privacy seems more endangered and therefore more precious.

Thus, it has been in the age of technology that the right of privacy has come to be defined legally. But the risks of excessive demands for privacy remain dangerous for democratic decision making, especially when modern technologies present risks that cannot be resolved through private action alone. The emergent nature of modern technological risks reminds us that too much as well as too little privacy is threatening to liberty, since our liberty is at stake in finding solutions to those risks. Those solutions can only be found through collective, public effort and decision.

Rights of Speech and Association

The exercise of rights, according to most definitions explored earlier in this chapter, presumes both speech in the sense of claiming and association in the sense of a sharing of moral rules. Thus, it seems a logical implication for any system of rights that individuals have rights of speech and association. It is indeed difficult to imagine a system of rights without these, and this says something important about the nature of rights that is overlooked or misconstrued by some theorists who criticize rights as being too "privatizing" or anticommunal. After all, the right to speak freely is not, as Jeremy Waldron points out, "the liberty to talk to the bathroom mirror," nor is the right of association the freedom "to prefer one's own company." Rather, as even Marx understood, such rights "are exercised only in the community of other men."[42] Rights exist as a medium for social intercourse.

In some important ways, then, the intertwined rights to free speech and free association are often said to imply each other. Virginia Held includes the right to form and join associations as one of five aspects of the right of free expression, along with the right to assemble and protest.[43] Her reasons for bringing these rights together recall the fears of John Stuart Mill and Alexis de Tocqueville that too much conformity in thought, speech, or action threatened the liberty that Americans professed to hold dear. Speaking ideas freely and joining with others who might hold the same unpopular ideas we do are essential liberties in a free society, according to Mill, and they are jeopardized together when either of two trends becomes manifest in society. The first appears when excess homogenization occurs in society and diversity is lost; the second, reminiscent of Madison, occurs when society fragments excessively into different and increasingly isolated "identity" groups. The second phenomenon leads to what today is called "identity politics," a fallen form of liberal pluralism anticipated by Madison and to which modern risk contributes much. I focus on it at length in chapter six. What is significant for our discussion here is that modern technology has given rise to both trends, and in the process new threats to the rights of expression and association have arisen.

These technological impacts are not obvious. At first glance it would seem that modern technologies have been an extraordinary boon to both rights of expression and association. Earlier in this century the inventions of radio and television signaled entirely new media for the expression and dissemination of ideas. With the more recent emergence of cable television and satellite transmission, the extraordinary potential of television as a medium of opportunity seems finally about to be realized, as the national networks watch their grip on the viewing public weaken and public access stations offer air time to almost all comers. Radio is both an older technology and one changed less by technical advances, yet the relatively sudden political significance of "talk radio" in the early 1990s has seemingly renewed that medium's contribution to the free exchange of opinion. And finally, the "information highway" with its worldwide network of computer-connected "cyber-speakers" is both less regulated than either television or radio and more wide open to the utterances of any subscribing member.

Clearly modern communications technologies have expanded the forms of expression and association available to any individual, and in that sense have altered the substance of those rights, but it is less apparent that the effect has been altogether liberating. With increased opportunities to speak and be heard both on television and the Internet have come new pressures to conform to the ideas of others, whether those others refer to all of society or some subset of it. Cable television, for example, does offer a wide variety of channels, over which the federal government has relinquished most of its regulatory powers. Yet, as Virginia Held makes clear, the opportunities for free expression opened by the new technology are "often completely overwhelmed by commercial interests and almost invariably depend on possessing, already, the economic means of expression." Indeed, as for the hope that more channels will mean new ways to challenge and even oppose the economic powers-that-be who rule popular culture, "the business society is remarkably adept at turning critics into just another product or service to be sold or used."[44] The culture of modern technological society still threatens to overwhelm creativity and free expression as Mill and de Tocqueville feared, in spite of the remarkable degree of individual access to the means of mass communication that cable and computer technology together make possible. The problem is, indeed, as Bruce Springsteen sang, that there are "fifty-seven channels and nothin' on," since "what's on" does not really change no matter the number of stations.

New communication technologies also feed into a growing feature of life in Western liberal societies known as "identity politics." Perhaps as a response to the anomic effects of Arendt's dreaded "mass culture," identity politics is an attempt to regain a sense of self through identification with like-minded (or -colored, or -gendered, or -raced, or -religioned, and on and on) others, who are then differentiated from everyone else in terms of their identifiable place within the larger culture. Within a melting pot culture such as ours, identity politics is nothing new, but the resistance to assimilation or identification with the interests and goals of others so apparent in US politics today revives the fears of the founders about the deleterious effects of faction on everyone's individual rights.

By offering access stations to almost all comers, a sympathetic ear on the radio or a kindred spirit somewhere out there on the Internet, modern communication technologies allow political discourse to retreat to the privacy of one's own living room, darkened study, or glowing computer screen. From the privacy of one's own home one can reach out to whatever kindred souls occupy cyberspace or radio space and vent one's spleen, relaxed in the knowledge that "responsibility for the views expressed here" will never be assigned publicly. These technologies offer a brave new world of unfettered speech and association, but it is neither truly public speech nor communal association because their uses presume none of the openness, and therefore public accountability, of democratic citizenship.[45] Rather, these are new technological opportunities for the growth of faction, and its effects remain as dangerous to free democratic exchange and decision as when first identified in 1787.

From this brief look at the transformative effects of new technologies upon four traditional rights, we can draw two conclusions. First, medical, data processing, communication, and energy technologies have altered the meaning or substance of our rights in significant ways. In some cases, as in the impact of medical and biomedical technologies on the definition of life, these technologies have reopened disputes about topics relating to rights that once seemed settled but are again open to question and on which important political consequences ride. So the abortion and euthanasia debates rage because of technology's impact on the language of rights, yet within that dialect a political solution to either dispute seems out of reach. In other cases, modern technologies promote withdrawal from the public sphere altogether into our expansive webs of private communication and entertainment possibilities. There our speech, privacy, and associative rights become the armor behind which our public selves remain hidden behind anonymous calls to the radio station or symbol-laden access codes on road signs along the information highway.

Despite (or perhaps because of) the new technology's heightening of the privatizing effect of rights, a second conclusion to be drawn is that the reliance on claims of right has not diminished within contemporary politics. Even though no longer necessarily (or even remotely) an impetus to social engagement and political discussion—if indeed they ever were—the pressing of rights-based claims continues unabated within liberal politics. This is certainly not a surprise, yet it is troubling precisely because decisions that need to be made about the risks of new technologies cannot always be made on the basis of rights or without infringing somebody's rights for the good of the whole. Risks call attention to our collective life; rights, particularly as affected by new technologies, allow us to retreat from it. But decisions for coping with the risks emanating from those technologies cannot be made from within the cocoon of individual rights, a private sanctuary created to some extent by the technologies themselves.

Rights, Interests, and Technological Risk

Our discussion of the interplay of rights and modern technology has so far led to the ironic conclusion that the risks generated from new technologies require collective or communal decisions even as they make avoidance of collective responsibility more possible for many individuals. They do so by heightening the efficacy of rights-based claims about what the community is *not* allowed to impose on individuals—e.g., the risks arising from proximity to waste dumps, restrictions on what one can say on the radio or on the internet, or the costs of medical treatments for someone else. Rights as presently practiced effectively "trump" policy decisions that require individuals to take the welfare of others into account, especially if some sacrifice of self-interest is implied. "Rights language," says Michael Ignatieff, "offers a rich vernacular for the claims an individual may make on or against the collectivity, but it is relatively impoverished as a means of expressing individuals needs *for* the collectivity" (emphasis Ignatieff's).[46]

Individually we have a need for collective decisions regarding the risks of new technologies—one might even argue that we individually have a right to demand solutions. But arriving at solutions will require sacrifice of interests liberal citizens are accustomed to assuming as a matter of right, whether they involve a long-term risk of irradiation or short-term invasion of privacy. It is this conception of interest that muddles policy disputes concerning, for instance, siting issues or distributions of risk. Private interests in such disputes that citizens view as essential quickly turn (in their eyes) into rights, and rights into trumps, and as a result the policy decisions never get made. The pervasive question for liberal society in the age of emergent technological risk then is simply this: can rights talk arrive at collective solutions for risk?

In her extended discussion of rights talk, Mary Ann Glendon suggests that in its excessive reliance upon rights, the liberal moral equation leaves insufficient room for "the missing language of responsibility" as well as the "missing dimension of sociality."[47] If she is correct, these absences within liberal moral theory do not bode well for coping with the emergent risks of modern technology. Indeed, if liberalism really lacks both conceptions, arguably it contains no moral theory whatsoever. That possibility will be taken up in the next chapter, where we will assess the adequacy of rights as a basis for either moral responsibility or political decision making.

The issue confronting us here concerns the relation between rights and interests, because in public disputes about the distribution of risks, citizens who feel that their private interests are being threatened often claim that their rights are being violated. If the claim is valid and the substitution of rights for interests is allowed to proceed, then in a society such as ours that acknowledges the inalienable rights of its citizens, the collective interest must give way. Thus, for example, after 20 years of site evaluation, selection, and controversy, in 1998 there is yet no certain future location for the disposal of radioactive wastes.[48] The politics of rights defined as coterminous with private interests has trumped every potential site, to the detriment of the future interests of all US citizens.

In issues like these, the question to ask is whether any individual or group really possesses an unshakable right to deny the public interest in siting a facility. The NIMBY attitude proceeds on a specific understanding of the relationship between interests and rights and a view of rights themselves as trumps. Earlier in this chapter we saw how most rights theorists assume that if there are such things as natural, human, or inalienable rights, they are tied to interests that are "essential" to the individual in some sense, in Raz's terms, or involve an object which it would always be supererogatory for the individual to relinquish, in Meyers's terminology. This relation between interests and rights is controversial on both logical and substantive grounds. Logically, as Hume pointed out, normative conclusions—rights-based prohibitions, for example—do not automatically follow from factual premises about human interests or needs. Substantively, it is not clear why rights should give priority to interests (or needs or attributes) that all individuals have in common instead of to "more restrictive, inegalitarian characteristics that pertain only to some persons, or to persons in varying degrees, such as expert knowledge or will to power or productive ability."[49]

Meyers's answer to these criticisms invokes the human need for moral systems and, concomitantly, any moral system's need to protect the lives and welfare of moral agents. That is a clever and successful argument, but, as Meyers recognizes, it only legitimates a very small number of inalienable rights, one of which emphatically is *not* the right to avoid all risk emanating from possible collective decisions.[50] One might indeed have an *interest* in avoiding such risks, but this does not qualify in and of itself as an *essential* interest thereby rendering it status as a *right*. And even if it did so, why would we necessarily presume that as a right it could automatically trump the collective interest in site selection? We need to explore both the concept of interest and its relationship to rights.

Substantial literature debates the meaning of the concept of self-interest, not all of which I want to summarize here.[51] As Madison understood, the variability of private interests, and the intensity of feeling often attached to them, made their protection a political necessity since they are the sine qua non of both faction and liberty. Interests arise from wants, as Brian Barry argues, but they are not therefore merely "want-regarding."[52] Interests also lead individuals to make claims publicly concerning them, many of which involve claims of right. Therefore, in opposition to Barry, S. I. Benn suggests that the idea of an individual interest is "very much like the idea of a natural right."[53] For Americans steeped in a deepening mire of rights talk, it is Benn's understanding of the concept of interest that legitimizes their NIMBY ways. This is not to accuse all Americans of simplistically viewing any or all of their "wants" as rights. If they did so, politics of any kind would be impossible, and Americans still understand the need for compromise. Nevertheless, I am arguing that it is in their exaggeration or misidentification of private interests as their essential interests that leads Americans into the miasma of NIMBY politics.

But though giving rise to the NIMBY syndrome, the politics of technological risk can also potentially point the way out of this terminological thicket surrounding interests and rights. More important, it can provide a way to resolve the conflict between protecting legitimate interests and rights of individuals, while still making difficult choices involving the collective interest. Since modern risks are emergent, they call attention to the inseparability of individual interests as they relate to the products of high technology. We all share an interest in minimizing the risks to which we are subjected, but it is not in our power to do so alone, given the nature of the risks we individually face. Our individual risk is not separable from that of others; therefore, it is logical to interpret our interest in avoiding risk as not individuated or entirely private either. In other words, as emergent risks confront our sense of our own self-interest, they force us to recognize that a growing part of our interests is what William Connolly calls our "social interests."

Social interests are what make social life possible; they presume an individual interest in maintaining good relations with other people. They are "the interests a person has as a social being" that "involve a reference to other people."[54] Such interests often require what seems to be a sacrifice of individual interest, but in doing so social interests elevate the whole concept of interest beyond its basis merely in private wants. Social interests *are* self-interested, but Connolly takes the "self" here as intensely situated in society—more intensely so than in want-regarding approaches like Barry's, or even Madison's, for that matter. Connolly explains, as follows:

> For instance, a person might decide to forgo a certain material benefit in the interests of maintaining a relationship of trust with another. He might simply cherish for itself the kind of life implied by sustaining such relationships more than he prizes the sacrificed material benefits. He is surely acting in his own interest in this case. There is *sacrifice* here, of course, but not in the sense that an act of altruism is the sacrifice of one's own interest. The person is simply sacrificing a lower-order interest for the higher-order interest he has as a social being.[55] (emphasis Connolly's)

It is essential for our dealing with the risks of modern technology that when citizens confront emergent risks, they recognize their individual interests as their social interests.

The concept of social interests also points up the disutility for risk decisions of Barry's or Benn's characterizations of interests as either merely wants or more grandly as rights. The NIMBY syndrome displays liberal citizens parading their wants as rights in the name of pursing their interests. But these are not their interests as social creatures situated within a context in which a difficult collective decision must be made—i.e., these are not their social interests. They are also not what Feinberg, Raz, and others call essential interests, the interests that stand behind legitimate rights. This is an important—indeed, the crucial—point, since if the interest in refusing to accept the additional risk that emanates from proximity to a needed installation is *not* an essential interest, then the NIMBY claim is invalid because rights are not at issue.

But why are the private interests of individuals at issue in NIMBY claims not essential when compared to their social interests, and therefore not the basis of legitimate rights claims? This is a loaded question, since if my position allows all NIMBY claims to be dismissed out of hand, it would seem the fate of all rights is then in jeopardy when a social need requires their abrogation. My answer is that legitimate social needs can never require denial of certain rights and that these (but only these) are legitimate rights because they are grounded on essential interests. They are not, however, the rights (or interests) at issue in NIMBY disputes.

NIMBY claims proceed on the basis of alleged violation of rights held by individuals regarding their property or health or those of loved ones. Property and health here count as the private interests involved. My contention is that although they are important wants, they are not in this context essential interests in that they are derivative from prior interests at stake in the same dispute. Those prior interests are of Connolly's "social interest" variety, and they take precedence as essential interests. Thus, they grant certain rights but not the right to say "not here" or, in other words, to win the NIMBY battle.

In general, emergent technological risks render certain social interests essential, at the same time that they deny the right-granting power of nonsocial, usually want-regarding, interests. This is an important consequence for moral and political theory, for it effectively changes the substance of rights within liberal societies. Property and health interests of individuals involved in siting disputes are themselves derivative from a prior interest of theirs—and shared with all others in the community—to site an installation that will remedy a risk emerging from their shared reliance on modern technologies. Why is this interest prior? It is so for three reasons. First, it is an interest that results from the previous, collective decision of society at large to rely upon technologies that produce risks at the same time they produce food, electricity, wealth, etc. Even though such a decision may have been an emergent as well as a collective one (in that there may have been no moment of conscious decision to accept the risks and benefits by either individuals or the society as a whole), it forms the initial context within which the impact upon other subsequent, private interests can be felt. In other words, although individuals might feel their interests threatened by the prospect of a generating plant, their social interest in electricity must be seen as requiring some responsibility on their part in accepting the costs of satisfying such an essential interest. We explore in the next chapter the extent of collective and individual responsibility in such cases. I emphasize here that the previous collective decision to rely on electricity is both temporally and ethically prior to whatever interests individuals might feel are threatened by its usage.

Second, the social interests at stake in a siting dispute are prior to interests claimed as rights by individuals because the social interest in averting the dangers of pollution or radioactive poisoning will remain even if another site is chosen. Thus, if some citizens succeed in moving the site elsewhere, their original interest in finding a disposal site remains unsatisfied. Third, averting these emergent risks cannot be achieved by individuals acting alone, yet if they are not averted future private action by individuals (such as staying either warm or healthy) may well be jeopardized.

Simply put, the emergent risks of modern technology transform the truly essential interests of individuals, whether in life, liberty, privacy, or expression, into interests that cannot be pursued in isolation. Therefore, the rights that entail from these essential interests are themselves of a distinctly social, interactive nature. The technologies have altered the interests and also therefore the substance of the rights emanating from them. Thus, these interests are in their essence social interests, not private self-interests; they can, as Connolly observes, include sacrifice, but "not in the sense that an act of altruism is the sacrifice of one's own interest." Social interests, as a consequence of the interpersonalizing effect of emergent risks, are our mutual, essential, "higher-order interests." As such they are the progenitors of a limited number of natural, inalienable, moral rights, one of which is *not* the right to obstruct the social interest in doing what is necessary to cope—collectively, of course—with technological risk.

What rights, then, are left to the individual once the risks of modern technology have produced their effects on individual interests? Only those rights that make individuals' knowledge of their social interest possible. That is not as enigmatic as it sounds, coming directly from Rousseau and Habermas. Even though often construed otherwise by contemporary commentators and legal practices, rights according to the theorists consulted early in this chapter are by nature concepts requiring a social context. On reflection that seems axiomatic, but, as Glendon's book chronicles in great detail, today rights are most often used to deny or somehow evade the whole idea of sociality and its incumbent individual responsibility for the common good. Thus, among other consequences, contemporary interpretations of rights have stood in the way of effective policy making for technological risk.

Still, rights and their exercise must be honored in any democratically designed decision making for risk, even if their number is limited only to those rights warranted by the true social interests of individuals. What are those rights? Essentially, they are what they have always been: the rights to whatever is necessary to exercise one's capacity as a human being to make decisions that affect oneself. The capacity for rights as elaborated by Raz, Feinberg, and others defines what rights are undeniable to individuals: whatever is necessary to make claims. Briefly, these include the right to speak freely in the public realm, to make one's case, to have it be heard, to have it—and oneself—be respected in the course of social decision making. What rights are not included? In disputes involving the public good, the right of any individual to win.

The rights properly belonging to liberal citizens are, in short, whatever rights necessary to make democracy possible: the rights to present one's interests and have them equally considered. The logic of this conclusion follows the form of Meyers's argument that legitimate rights are those necessary to make moral systems possible. A more frankly political view of rights than hers, it presumes that rights are both "contextual" and "discursive" in McGinn's and Chambers's usage. They are contextual in the sense that modern technology and its attendant risks have changed the context of democratic life. This does not mean, as people sometimes assume by saying that something is contextual, that rights are completely relative or contingent. Rather, rights are relative to the context of democracy, a context altered by modern risk. But democracy always has specific and absolute requirements concerning the protection and respecting of individuals. Technology with its risks does not change the fact that there can be no democracy if individuals do not possess the rights to be heard and to have their interests treated equally with those of others.

The rights that modern risks leave for individuals are also distinctly discursive. As evident by now, the rights to be heard and respected presume a discourse within which those reactions can occur. Such a view of rights will necessarily alter the process of decisionmaking for risk, as chapter six explores at length, because at present the policy process for risk does not guarantee such a reception for individual interests. The discourse defined by the essential interests and rights of citizens within the modern technological community must be a vastly more participatory one. Without increased consideration of individual interests in the policy process, the essential interests of society as a whole will be jeopardized. Furthermore, because rights by themselves will no longer act as trumps, protecting individuals and minorities from abuses of power, citizens must avail themselves of their ante of participation in furtherance of their interests.

Such consideration and participation may not generate the level of concern one might expect in more communal settings such as a family, but it does suggest at least the necessity of cooperation. Richard Flathman says that "rights provide the individual with at least some of the elements of a *place*, an *identity*, a role in the *social* milieu" (emphases Flathman's). Although relationships defined by such elements involve "a certain holding back, a certain reserve," as Flathman notes, they are not entirely self-interested or egoistic:

> Rights involve a certain holding back, a certain reserve, in relationships with other people. The participants in the practice are explicitly and in very substantial respects parties to such relationships, but they enter into them, as it were, on condition. An area of action remains with the participant. Although the relationships of which the practice consists are cooperative in vital respects, they are also, at least typically, significantly competitive.[56]

The practice of rights is a cooperative and discursive enterprise. It may be, as Flathman recognizes, an enterprise to which participants do not give themselves over fully, but nevertheless they must recognize that within an area of mutually (albeit competitively) defined and shared interest, their right is to participate, to make and hear claims, not to end the relationship or the competitive practice itself through the "trumping" action of their rights.

Flathman's understanding of the "practice" of rights is useful in viewing the scope and limitations of rights-based claims involving the risks of modern technology. The risks of modern technology may be egalitarian in that ultimately they affect everyone, but they pose real authoritarian dangers in the temptation they pose to those in authority to ignore the real rights of citizens who either individually or as part of a minority stand to bear a larger burden of cost in coping with the risks. This is why democracy is essential as a counterweight to risk. The democratic process for risk policy must include the rights of all citizens to present their interests and claims regarding the presence of risks. They must be allowed into the game—the discourse—of decision making. Democracy protects the individual's rights to be heard and to be respected as the cornerstones of liberty and equality. This protection ensures that in the decision to distribute risks, all citizens have an equal right: to have their interests receive a respectful hearing. Their rights are their entree—their ante—into the competitive game of decision making under risk, known as risk policy.

Conclusion: Rights, Risks, and Democracy

The ante of rights properly bestows upon individual citizens special participatory opportunities as well as responsibilities within an acceptable and successful risk policy process. As that process is organized with an eye to protection of citizens' legitimate rights, it emphatically endorses the necessity of democracy. Without democracy there is no protection for the individual chosen to suffer the dangers of proximity to risky installations or practices deemed essential to the common interest. In this sense nondemocratic societies have considerably fewer problems in dealing with risk—without the assumption of rights as antes, siting disputes would simply never occur. Nevertheless, modern risks do challenge democracy's principles by the mere fact that absolute equality of treatment is impossible when it comes to the siting of certain installations. Somebody will live closer to these sites than others; thus, their liberties will be constrained more than those of citizens who live a distance away. This inequity is inevitable, but democracy guarantees that equality can yet prevail, as long as we recognize that the existence of risk demands that we view our interests and rights as part of a transformed social discourse and policy process.

The sixth chapter explores how that discourse and process differ from what exists today. But here I conclude by acknowledging how the emergent risks of technology have brought us, if not a new, then at least a recalled, almost forgotten understanding of the nature of individual rights. Bentham once warned of how, if invoked improperly, rights and rights talk can become the terrorist language of self-concern that threatens the public good. His image resonates within our political reality today. We are acquainted with the face of terrorism in the late twentieth century but, as we will see in the next chapter, just beginning to recognize its various expressions.

If wrongly construed but successfully pressed, rights *can* have the impact of terrorism when wedded to the potential for catastrophe that some technological risks embody. For example, as the Cold War winds down and the glut of weapons-grade plutonium and uranium escalates due to decommissioning and dismantling of nuclear weapons and factories,[57] the utter lack of a disposal policy for radioactive materials looms as chillingly as ever did the specters of armed missiles in their silos or of terrorists' bombs. The policy absence here is a direct result of the rights-as-trumps capability of citizens in localities identified as prospective neighbors to disposal sites to forestall decisions. As long as rights continue to mean the ability to ignore shared collective responsibilities—to win policy disputes rather than only to be heard respectfully during them—technological risks will befuddle liberal governments with potentially disastrous consequences.

Emergent risks require, if not a completely new understanding of rights, at least a renewed sociality in their application. This is a hard lesson for our litigious times and one that necessarily vitiates the individualism of liberal theory. But liberalism has never spoken with one voice on the issue of rights or on other concepts surrounding the ideas of responsibility and moral obligation for individuals or, increasingly, for groups. This is the next stage of our discussion—the impact of technological risks on liberal notions of moral behavior and ethics. Rights under some definitions have certainly always been a part of liberal moral theory. But rights cannot remain liberalism's entire source of moral duty and action. In their emergent status, technological risks lead us to renewed discussions of what groups and individuals owe each other as a matter of obligation, if not exactly of right.

5

Risk and Responsibility

Jeremy Waldron's exploration of rights concludes that, in terms of providing a complete moral theory, no theory of rights is capable of standing on its own. Many human relationships, including those among families and between intimate friends, are easily destroyed if some or all partners become too concerned with entitlements due them from the others involved. These relationships demand something other—and more—than mere respect for rights, and it is the responsibility of all involved to recognize what more is needed for a healthy relationship. Somewhat hopefully, Waldron speculates that perhaps the role of rights in such communal webs of reciprocity is to clear some space for "each individual to take upon himself the responsibility for molding and constructing a life with others and of providing the social and spiritual conditions that make that responsibility possible."[1]

Waldron's optimism about the effect of rights in the acceptance of responsibility is perhaps itself born of the traditional assumption within modern moral philosophy that everyone is particularly liable for the welfare of those closest to him or her. Responsibility and moral obligation, most moral philosophers since Hume have agreed, exert the strongest moral claim on us within our close relationships with those we know and love and diminish as the focus moves outward from family to neighborhood, town, nation, species, or ecosystem. In any case, Waldron's hopefulness is bolstered by liberalism's historic faith in each individual's capacity for moral autonomy and therefore responsibility, attributes that qualify each person for the benefits of liberty and justice. Even though moral responsibility is not a particularly individualistic concept when compared to some others, it nevertheless plays an essential part in liberal individualism's defense of its primary values of liberty and justice.

The notion of responsibility, according to Larry May, "has more to do with how people fit into a larger social context than it does with what individuals owe to each other as their due (the way justice is traditionally characterized)."[2] This explains its intrinsically social rather than individualistic character, but responsibility nevertheless is an essential component of how liberalism views individuals in the world. As much as any other liberal principle we have considered—consent, authority, rights, among others—the attribution of moral responsibility to the individual for his or her actions is the foundation upon which liberalism makes its political and moral appeal.

We have already seen how technological development and the emergent risks it generates upend the original meanings of many foundational liberal concepts. Chapter one explored how the realm of politics is widened by the togetherness that risk forces on the private world of individual citizens. In chapter two the weakness of liberalism's view of consent was revealed by its inclusion in models of risk assessment. In addition, in the absence of a strong obligation to community within liberal theory, we observed how liberalism's subsequent inability to resolve risk issues challenges its view of political obligation as well. Chapter three exposed the vacuity of liberalism's understanding of authority. Often commented on by political theorists, this void never appears more clearly than in issues of risk policy that cry out for an authoritative moral voice to legitimate sacrifice and to transcend the narrow legalisms of a judicial authority mired in a litigious explosion of liability claims. Finally, chapter four recounted the ebbing power of natural rights as liberalism's moral compass. Not only do modern risks disconcertingly redefine the substance of basic rights, but the political machinery designed to incorporate rights claims into successful risk policy has ground to a halt, beached amid a sandstorm of rights-based NIMBY claims.

Underlying each of the arguments made in earlier chapters is a perception that somehow modern risks are redefining how individuals must relate to each other, not only for debatable moral or political reasons, but because our continued existence—either political or material—depends on it. Furthermore, in light of the argument so far, it is increasingly obvious that liberalism's tired conceptual nostrums are part of the problem that emergent risks elucidate, not part of the solution. The risks we face today are unanticipated by liberalism's individualist epistemology; thus, the emergent collectivism of modern life cannot be reduced to the individualistic equations of liberal political and moral theory.

Central to any theory of how to cope with the risks of modern social life—either a failed liberal theory or a potentially successful new one—must be a notion of responsibility. Risks portend harm for those subjected to them; thus, they raise traditional as well as new questions for moral theory. Who should be held responsible for the riskiness of modern life—that is, who caused it? Or, perhaps more to the point, who bears the responsibility for ameliorating it—whose task is it to make life safer? Furthermore, what should be the distribution of harm and responsibility within a technological society beset by the emergent risks of modern life—that is what are our moral obligations relating to risk? These questions occupy this chapter. They make clear the varied meanings attached to the term responsibility within common discourse, what I will call, following Robert Goodin, causal, task, and moral responsibility.[3]

Liberal theory has for 200 years postulated a theory of moral responsibility based on tenets concerning causality, free will, and individual choice. This chapter argues that the emergent nature of modern risk seriously challenges the efficacy of all three elements. As a consequence, risk threatens to leave liberalism bereft of a basis in moral argument to guide the distribution of responsibility necessary for successful policy making for risk.

Beginning with this chapter and continuing into the next, I also try to demonstrate how viewing risk philosophically requires that several apparently disparate literatures (or discourses) must be combined. Of course, such an effort has been part of earlier chapters as well. In fact, an underlying theme of the book has been that policy making for risk requires that we take seriously research from a broad spectrum of intellectual disciplines. But more than this, I have wanted to show in this book that it is important for philosophers, political theorists, and other toilers in the fields of conceptual analysis to realize that when they focus their work ostensibly on ideas such as rights, consent, authority, or the like, they are also writing about something as apparently down-to-earth, even mundane, as living with risk. So I refer to numerous articles, books, and authors in the pages that follow, from several different disciplines and even from news reports presenting the "discourse" of everyday life. In some cases I in fact rely on the arguments provided in these works, others I dispute, but I include them in order to bring risk into the world of the philosopher, to bring philosophy to risk, and generally to attempt a rapprochement between theory and everyday life.

In this chapter I review some contemporary moral theories that offer interesting new conceptions of responsibility that might be helpful in redefining our obligations to each other in the face of our risky present. Prominent among these are the arguments concerning collective responsibility by Larry May, Peter French, and political scientist Dennis Thompson. In the last section of the chapter and the conclusion I also adopt some of the arguments of Robert Goodin and Marion Smiley concerning the need for a new, consequences-based rendering of moral responsibility that omits blame as part of the definition of responsibility. We explore these offerings as we discover how traditional renderings of responsibility within modern moral philosophy and liberal theory are of little use when we confront the dilemmas of coping with risk.

The following discussion of responsibility continues to rely to an extent, as in previous chapters, on some customary methods of argumentation within moral philosophy and political theory. However, the challenge that risk poses to traditional understandings of harm and responsibility requires a second, different approach as well. Following Virginia Held,[4] I believe that contemporary issues of responsibility demand that we adopt a less metaphoric language of moral discourse. This language is most evident in the hypothetical examples of moral dilemmas philosophers use in their arguments for the meaning of significant moral terms like responsibility. The problem pervades philosophical discussions of many concepts of morals and politics, but the topic of responsibility seems particularly prone to this approach.[5] Thus, for instance, Hobbes has his state of nature, Rawls his original position, Nozick his "experience machine," and many contemporary philosophers' ubiquitous "desert island." The arguments in favor of such abstract models are well-known—they are meant to separate out any self-interest or impartiality of our own as we decide what morality demands in a given instance. The danger of such methods is that they hide the often specific, capricious, and uncontrollable nature of harm as it presents itself in the real events of our lives, as well as failing to reveal the deeply hidden prejudices born of self-interest but not recognized as such. Modern emergent risks as we have discussed them embody both these dangers. As emergent, their causes often are not easily identifiable, and decisions about their acceptability often presume or perpetuate political inequities. As Held points out, "to take the point of view of the ideal observer is to abandon the point of view in which our experience is real."[6] In any case, there is no need to rely on metaphor to explore the moral conundrums presented by modern risks. Real life offers examples aplenty.

As we experience the real, emergent risks of modern technology and the moral puzzles they present, it is essential that we not "lose" their reality in attempting to model a method to resolve them. These risks require political solutions, and politics is a realm where the clash of specific interests is not easily generalizable into an "ideal theory" or metaphorical approach to the normative questions (such as where responsibility lies) contained within them. Held describes the dilemma of how to make moral theory relevant to the political world of clashing interests and emerging risks:

> Moral problems often arise in political contexts and need solutions within such contexts, solutions that are justifiable in a normative and not merely instrumental sense. But political contexts are not even describable without reference to the actual interests of actual individuals and groups. Yet such interests find little place in ideal theory [such as, for instance, in Rawls's]. In actual political contexts, moral issues arise in questions of maintaining or furthering or harming actual interests, and deciding between actual interests in the almost ever-present situation of conflict between them. We cannot hope to deal with such questions in a real political context without considering the actual interests being felt, asserted, claimed, or at least thought about in them.[7]

Held's admonition should, I believe, be taken seriously throughout all of moral theory but especially as moral philosophers and political theorists confront the issues of harm and responsibility contained in the politics of emergent technological risks. To assist us in this regard, I propose to explore the issues of risk, harm, and responsibility manifested in a horrific experience shared by millions in Spring 1995. Rather than a hypothetical example on which to ground the subsequent theory of responsibility that emerges in the rest of this chapter, the terrorist bombing of the Alfred R. Murrah Federal Building in Oklahoma City is an actual experience in which, in one way or another, most Americans and television watchers over the world participated.

There are benefits as well as dangers in this news-related approach to moral and philosophical investigation. A prominent drawback is the temptation to rely overly much on the engaged sympathies of citizens still horrified by the event. Though my approach to the issue of responsibility will rely on what Strawson calls "moral sentiments,"[8] I am well aware that neither his nor earlier renderings of such elements of human psychology (such as Hume's or even Plato's) confuse sentiment with sentimentality. Further dangers posed by such an approach include relying on the sensationalism of the event or assuming erroneously that the event remains fresh enough in readers' minds to make the point. I will scrupulously endeavor to avoid the former but, perhaps somewhat optimistically, continue to assume the latter.

Because shortly after the bombing suspects were identified and arrested, this example might seem out of place in our discussion. In this example, the issue of avoiding the risks of such an event is clearly not merely one of collective choice, because it also presents an actual crime and anticipated punishment for the perpetrator(s). Thus, it might seem more appropriate for us to explore the idea of responsibility using issues attached to the modern technologies of energy, communications, medicine, and the like—examples of which abound in previous chapters. I am aware of the difference between this example and others I have employed, but I discount it for three reasons. First, plenty of other examples of risk and responsibility have formed the basis of philosophical discussion that, if they did not include elements or charges of criminality, probably should have. Prominent among these are the cases of industrial malfeasance at Love Canal and of Union Carbide in Bhopal. The horrific fruition of technology's risks in these instances raises responsibility issues of the most fundamental sort as well as speculations of what should, in fact, be considered criminal activity or omission.[9] In all of these cases, furthermore, resolving the criminal issues potentially at play in them does not make the moral issues any less pressing.

Second, moral philosophy must find its voice in instances like the Oklahoma City bombing because, as we will see in the next section, within that event are contained several different kinds of risks, all of which present moral consequences of interest to contemporary ethical theory. In avoiding real examples such as this one, philosophy and philosophers risk missing much of the challenge that modern technological risks pose to philosophy and to society as a whole. That challenge is intrinsically moral and extends not only to potential criminals but to law-abiding citizens as well. In other words, there is more responsibility to be distributed for Oklahoma City than that borne by the perpetrators of the crime committed there.

Finally, using examples like the Oklahoma City bombing recall—almost weirdly and certainly frighteningly—Bentham's drawing of the connection between moral concepts in calling rights talk "terrorist language." This almost sounds too pat; nevertheless, it is significant that the alleged perpetrators of the bombing were motivated by what they perceived as challenges to their basic rights to bear arms and protect their property. Such motivations show the power of moral concepts (as they are elucidated by modern risk) in moving citizens to take action. It is impossible to understand the discourse within liberalism today concerning risk if we do not see how the risks of modern life can, on occasion, lead to violence. Modern technologies do make it easier for and to some extent more likely that government will encroach into the private realms of individual life. The previous chapters have shown that risk forces us together in fundamentally new ways that we can easily see might lead to violent reaction on the part of some. This is not to say that the fact of modern risk *caused* what happened at Oklahoma City, only that what happened there elucidates an aspect of modern risk worth exploring further.

In my focus on the terrorist bombing in Oklahoma City rather than on some arcane hypothetical examples, several features of modern risk and responsibility will emerge. First, we will see the insufficiency of traditional liberal and contemporary moral philosophical views of responsibility in explaining all the levels of risk and responsibility involved in this tragedy, especially the elements of collective responsibility. Second, by recalling shared memories of this event, we will naturally be drawn into a different approach to responsibility that Larry May terms "shared responsibility," that better describes the correct normative response to emergent risks. Third, we will see how the "authority of ends" posed by such emergent risks leads us to accept a consequentialist approach to issues of responsibility but, following a new interpretation by Robert Goodin, one that can also accommodate some valuable deontological contributions.[10] Finally, I admit that the use of real-world experiences to ground moral theory bespeaks a fundamentally pragmatic approach to ethics. I do not apologize for this, but I invoke Marion Smiley's recent sustained defense of pragmatism to show how emergent risks reinvigorate the fundamental truth enunciated first and best by John Dewey, that moral meanings arise from the "actual tensions, needs, troubles," confronting a community.[11]

One Disaster, Several Technologies, Emergent Responsibilities

A benefit of relying on hypothetical examples rather than actual events in presenting arguments in moral theory is that one can avoid sentimental, maudlin, or sensationalistic renderings of real life. We should endeavor to do the same in our employment of actual events, though as Hume recognized, we tend to feel the pricks of responsibility most sharply when confronting the real-life consequences of actions as they affect persons closest to us. In fact, even those moral theories most reliant on abstract reasoning and make-believe situations tend to confirm a differential level of responsibility contingent upon proximity, under some definition.[12] This moral "particularism," whether a part of utilitarian or deontological theories, depends for its appeal not only on cold rationality, of course, but on some appeal to human feeling or sentiments. For Strawson and ethical naturalists generally, this invocation of our sentiments is what makes morality possible and, for Strawson, allows him to keep pessimism at bay. It is also what makes responsibility to those closest to us seem most special as well as most strongly felt, as it is for them that we have the strongest emotional attachments.

In what follows, I argue that a prominent consequence of modern technologies and their emergent risks is the shrinking of our moral world—we are living "closer" to each other than ever before. Modern risks engage us in other people's lives—and they in ours—in ways and to an extent not possible until the age of modern technologies. Thus, in a very real sense, others to whom we are not related or near geographically or personally nevertheless engage our moral sentiments in a manner that should (and in some cases demonstrably does) lead us to accept moral responsibility for the harm having befallen them. At the same time, our sense of concern and responsibility for heretofore distant others is not the product of our having freely, autonomously *caused* the harm brought down upon them by the technologies or their attendant risks. In other words, the demands of proximity and causal efficacy requisite in most modern philosophical models of responsibility are clearly not being met, even though, as I argue, our responsibility nevertheless entails them. In short, the nature of modern risk has expanded greatly the number of persons for whom the moral particularist position contends we are responsible. Instead of presenting a hypothetical example of how modern risks accomplish this, let's bow to the undeniable impact of some of those technologies and consider the case of the Oklahoma City bombing of April 1995.

The tragic bombing of the Alfred R. Murrah Federal Building should be seen as the product of several different technologies, each of which carries a distinctive brand of risk and gives rise to different issues of responsibility. Some of these technologies are modern in the sense I have been referring to throughout this book and thus offer new types of emergent risks and special problems for the assigning of responsibility. Let's begin with the most obvious technology in evidence there, that of explosives.

The capacity to blow things up is not new to our age. Gunpowder has been in use for hundreds of years at least; thus, the first risk in evidence at the bombing of the Oklahoma City federal building—that of getting blown up—is not particularly unique to our time. True, the bomb in Oklahoma City, though easy to construct, employed chemicals found in the right combination only in contemporary fertilizers. Still, the risk of being victimized by a planned and controlled explosion has been all too present throughout much of human history for centuries.

Responsibility for the harm to victims in Oklahoma City by the (realized) risk of explosion is also relatively easy to assign. At this writing, only two alleged perpetrators have been found and arrested, but if others are found the process of determining guilt will follow a familiar pattern. The action certainly appears to have been intentional: that is, it was caused by the free actions of one or several whom the law considers morally autonomous agents. Responsibility issues here are relatively straightforward, since the technology and its risk we are addressing here is not of an emergent variety. Thus, the issue of causation is easily resolvable. Any moral theory would hold the bombers as the ones responsible because they clearly caused the harm that ensued.

This first technology in our exemplary event poses few problems, then, in determining the question of responsibility. Before moving on to the second technology present in an efficacious way in the tragedy, it is worth asking if the explosive technology employed were ultra-sophisticated and modern, would the issues of risk and responsibility remain as straightforward. Since the 1960s, psychologists and risk researchers have noted the different reactions subjects display when confronted with risks considered traditional compared with new and unusual ones or especially catastrophic ones. Thus, for instance, if the explosion in Oklahoma were the result of a nuclear weapon or from stolen fissionable material, or were accompanied by the spreading of a recently developed and deadly bacteriological weapon, would there be remaining issues of responsibility once the perpetrators themselves were caught, tried, and punished? And are the risks from such dangers in any way different from those accompanying explosions of fertilizer mixed with fuel oil (the technology of choice in Oklahoma City), or of gunpowder?

Questions such as these are difficult to answer precisely because more sophisticated technologies such as nuclear or bacteriological themselves include developmental histories encompassing many different people, government agencies, and public as well as secret decisions. That is, these technologies and their risks are emergent products of our technological age and thus also of collective actions not easily reducible ontologically to specific individual causes or agents. Therefore, the issue of responsibility is not completely reducible to the actual individuals viewed as perpetrators (the bombers) or, more philosophically, as the precipitating causal agents of the event. At the very least, we would expect investigations into the procedures and actions of agencies, corporations, and officials that might be held collectively responsible for the development or availability of the explosive material used. We might find then that additional responsibility needs to be assigned, still on the basis of cause, to agents more distantly removed from the event in terms of time. The purpose of such explorations would be to determine if some action (or inaction) on the part of other groups or individuals is also blameworthy and its agents therefore responsible. Attributing such additional responsibility would expand our view of responsibility somewhat but still maintain a direct connection between ideas of causation and responsibility.

These additional questions of collective responsibility are considerably more complex to answer, and they are raised at least potentially by the second technology present in the Oklahoma blast. This is the technology of modern construction techniques—the technology of skyscrapers. The risk emanating from the technological ability to house thousands of persons in one building is, I believe, different in kind from that of the first technology.[13] It is the risk of getting blown up *together* with a large number of fellow human beings.

What makes the quality of this risk different from the simple risk of explosion is its catastrophic nature. Psychologists point out that the difference is not merely one of quantity—between the death of hundreds or even thousands in one event rather than an equal number of casualties in hundreds or thousands of individual events. Rather, the difference lies in the perception that individual choice or even action is not a relevant factor either in the creation of the risk or in its possible avoidance. The risk is simply *there*—a pervasive feature of life more or less impossible to avoid and about which responsibility is extremely difficult to credit.[14]

Skyscraper construction technology is a twentieth-century knowledge, amassed slowly as the result of countless innovations in technique and building material. Living in modern society means that we are often in the presence of the technology—either by living and working in a tall building or merely by walking past one. Sharing at least occasional proximity to skyscrapers is not a matter of individual choice for most citizens in a technological society. Thus, if a skyscraper blows up—or if the technology fails for other reasons, natural (e.g., earthquake) or man-made—the question of who to hold responsible for the specific risk that comes from this building being a skyscraper rather than, say, a one-story building housing a dozen people, is different from that of who planted the bomb. It is different precisely because the issue of cause is so much more difficult to resolve.

Of course, the risk that a skyscraper rather than individual houses or offices will be blown up by terrorists may be quantitatively greater as well, because the scale of the destruction and its catastrophic nature make skyscrapers more appealing targets.[15] But, again, the responsibility for the existence of a skyscraper to blow up is, if it exists at all, a matter of collective choice. Thus, responsibility here has less to do with cause and more with the issue of who is charged with the task of deciding about the risk—either in the sense of determining if it will be allowed or in the sense of regulating it, once approved. This is a different sense of responsibility, one having to do not with cause but with the task of decision or regulation. What Goodin calls "task responsibility" is fundamentally different from causal responsibility and presumes no direct connection between the two.

Task responsibility is sometimes—and more often should be—an issue in local zoning decisions concerning the construction of skyscrapers. When zoning disputes do occur about construction, they usually center on aesthetic or crowding issues, but the risks that accrue from the concentration of people in any case should not be ignored. In areas subject to earthquake risks, this concern is nothing new and, in fact, has led to decisions resulting when citizen initiatives succeeded in blocking further skyscraper construction.[16] Task responsibility may possibly be assigned here—to corporations, patent officers, zoning boards, inventors, even all citizens—but they must be seen as agents of a different type of responsibility than what causal agency levies upon a lone bomber. Collective task responsibilities like these lie behind many modern technologies and the risks they present. We explore them fully in the next section.

The third technology in attendance at the bombing in Oklahoma City is the most definitive of our time and of its special, emergent risks. It covers the group of technologies generally referred to as "mass communications" instruments, including television, radio, and computer communication networks. These technologies played a variety of roles in the event itself and its aftermath, and several narratives of guilt or responsibility have already been entered by those with political or ideological axes to grind. The risks that these technologies present are the most emergent and inescapable of all; in many ways they define life in modern technological society, as well as its special susceptibility to terrorism. I discuss only two of these risks, the second of which is the most befuddling to traditional moral theories of responsibility.

First of all, it is entirely possible that the risk of this bombing ever occurring is itself partially a causal product of these three modern communications technologies. This is speculative, of course, but consider the three specific contributions these technologies made to the event. First, whether or not President Clinton was fair in implying after the bombing that "talk radio" should shoulder its share of responsibility, it is at least true that before the event there was considerable talk on various shortwave networks of "something big" in the works as a protest against government intrusion into the lives of individual citizens. Certainly, following the event, some radio hosts and broadcasters went overboard in justifying, in some cases even exculpating, the perpetrators. Second, the internet has long been a source of "how-to" information for those wishing to wreak incendiary havoc. Hours after the blast, in fact, an Internet bulletin board offered advice on how to build a "bigger and better" bomb than the one used in Oklahoma City. Third, it is television coverage of course that terrorists and other would-be bombers most covet for their actions, a predilection that goes far in explaining why such events are increasingly the tactic of choice for modern warfare.

To the extent that it makes sense to hold these technologies or their users responsible in some measure for the events in Oklahoma City, we should acknowledge two features of the concept of responsibility being applied. First, it is another instance of causal responsibility. Second, it is another instance of how modern technologies have broadened the conception of causal responsibility. These same technologies played a different role in the tragedy there as well, however, and thus introduce a version of responsibility different from either the causal or task responsibilities we have noted thus far.

In its ability to deliver visual and audio images of events as they happen, television presents a special second kind of risk felt by those "tuned-in" to events in Oklahoma City. Television makes it possible to vicariously experience the horror, terror, and grief that such events generate—to be "close" to those directly experiencing the harm itself. It is by now a hackneyed observation that television (like all communications advances to some extent) has shrunken our world to the size of the living room. The risk that such technological participation in distant events generates is mentioned less often. It is the risk of knowing, of being intimately caught up in another's experience of pain and grief. Such knowledge often seems to spawn a sense of responsibility to share in a more direct way with the actual victims. Thus, in response to the bombing, as in countless other cases of disasters also fully covered on the six o'clock news, citizens from around the country and the world sent their assistance to the survivors.

The risk of knowing is not unique to modern communication technologies, of course. Advanced medical, genetic, and reproductive technologies engender knowledge about our bodies or even of the fetuses within them that can be accompanied by frightful decisions that we otherwise would not be forced to make. These risks are such that patients usually must sign waivers absolving doctors, technicians, or other medical personnel of responsibility for the pain that the knowledge *itself* might cause. The recipients are required to be responsible for their own reactions. As another example, Henry Kissinger was fond of pointing out that even if all nuclear weapons were to be destroyed tomorrow, the knowledge of how to rebuild them would remain a terrifying threat to peace for the foreseeable future.

It can be argued that knowledge always brings such risks—that in the search for truth we risk discoveries that are at least unpleasant either spiritually or psychologically. But the knowledge that comes from the televised immediacy of experiencing another's pain is different still. It carries a special moral force because it brings us closer to those whose vulnerability and pain we sense to be, in some way, our responsibility simply because we are aware of it. Our televised viewing makes it difficult to maintain the distance between ourselves and the "other"—the victims. This distance, for most moral theories, absolves us of responsibility for the victims' plight; losing it brings us into a moral relation with them. David Campbell names this phenomenon of modern life and its technologies the "deterritorialization of responsibility."[17]

Employing concepts and arguments from the postmodern theories of Foucault, Derrida, and the earlier Emmanuel Levinas, Campbell argues that a postmodern understanding forces the recognition of additional "interdependencies" between what liberalism once assumed were autonomous and radically separate individual lives. Even though Campbell does not focus specifically on modern technologies and the risks they entail, he assumes them in his argument that a fundamental feature of postmodernity is the blurring of subjectivity—of the distinction of self from other. We have discussed this conclusion before in chapter one as one of the effects of the emergent nature of risk. Modern risks come from all of us together, emanating from our conscious and unconscious decisions to accept the benefits of modern technologies. Those decisions make us responsible for the risks that follow, both individually and, as will see, collectively.[18] They do so even if we can no longer assign causal responsibility for risk to any particular person, persons, or events. This responsibility begins with the realization that we are more intensely involved in each other's lives than liberal theory has historically assumed. We are, as Campbell says, "already ethically situated" in relations involving responsibility:

> This argument leads us to the recognition that "we" are always already ethically situated, so making judgments about conduct depends less on what sort of rules are invoked as regulations, and more on how the interdependence of our relations with others are appreciated. To repeat one of Levinas's key points: "Ethics redefines subjectivity as this heteronomous responsibility, in contrast to autonomous freedom."[19]

This surely is a different notion of responsibility. Although we have witnessed its acceptance countless times when citizens come to the aid of victims after innumerable televised disasters, we are reluctant to call it anything other than benevolence. I argue later in this chapter that such sharing in response to tragedy is in fact an obligation of modern life, a responsibility that the emergent status of our risks makes necessary through the diminishing of the moral space between us. This view of responsibility does not fit with traditional models of responsibility, blame, or obligation within most contemporary moral theory. However, just as we have recounted their effects in our explorations of contemporary views of consent, rights, and authority, emergent risks here also must alter our understanding of how responsibility is distributed in modern, technological society.

Risk and Collective Responsibility

The philosophical problem of responsibility posed by the emergent nature of technological risk lies in the difficulty in identifying individual agents related either causally or as a matter of task definition to the harm portended by risk. Most contemporary moral philosophers follow Joel Feinberg in assuming that responsibility presumes causal efficacy; that is, a person is responsible for harm if he or she knowingly and willingly *caused* it to happen. Of course, as our example of Oklahoma City demonstrates, causation is often controversial, because many persons and actions can ultimately be involved in actually bringing the harm to pass. But, according to Feinberg, for responsibility to be assigned, it must be possible to determine, first, causal contribution; second, that the individual responsible was in control of events—that is, there was no other external causation, randomness, or luck involved; and, third, that these two determinations can be "read off the facts or deduced from them; there can be no irreducible element of discretion for the judge."[20]

If these criteria are met, contemporary theory concludes, moral responsibility can be assigned and, for Feinberg, its difference from mere legal liability made clear. Moral responsibility is different because it is restricted to the "inner world of the mind, where the agent rules supreme and luck has no place; for here is a domain where things happen without the consent of uncooperative nature, where bodily movements are initiated, acts of volition undertaken, intentions formed, and feelings entertained."[21] Responsibility, then, is a consequence of control, choice, and independent volition, all of which can be ascertained as a matter of fact, and all of which only pertain to rational individuals, whether defined as human beings, or, by law, as persons, including "legal individuals" such as corporations.

Emergent risks make a hash of all these generally accepted requirements for responsibility. As we have seen throughout this book, modern risks resist reduction to individual actors and events; thus, they appear simply to "emerge" at the level of social life. We have been careful not to conclude from this feature that technology is itself "autonomous" in some mystical way in its ability to impose risks. Rather, to recognize emergent risks is to realize that individuals do not have control over all events that happen, at least when acting only as individuals. Emergent risks are the product of collective action, and thus coping with them is a matter of collective will. In other words, it is not always possible to find an individual causally related to some risks who can be blamed for them—some harms are the product of our life together and cannot be reduced further.[22] Thus, as Marion Smiley concludes, responsibility over such matters is not as purely factual a matter as Feinberg claims and depends "on a variety of social and political considerations which we either make ourselves or inherit from others in the form of social expectations."[23]

The risks issuing from the two technologies of skyscraper construction and television (or, broadly, modern mass communication) visible in the Oklahoma City bombing are emergent in nature. As such, they raise difficulties for the traditional view of responsibility restated by Feinberg from its original formulation within liberal individualist theory by Locke, Hobbes, and Kant. The skyscraper technology presents issues of what has come to be called "collective responsibility," which we explore in this section. Even less amenable to liberal individualist comprehension are the responsibility issues arising from the risks posed by television. These I call in the next section of this chapter questions of "shared responsibility," which should be distinguished from most models of collective responsibility and which move us even further from the traditional view of responsibility. The collective responsibility view examined in the next few pages is something of a middle position in that it is usually characterized in individualistic terms, either by focusing on the individuals within the collectivity (those "many hands," in Dennis Thompson's well-known argument) or on the collectivity itself as a fictional but legal individual (such as a corporation).

As of this writing, investigations into possible dereliction on the part of those involved in either the construction or security of the federal building in Oklahoma City have not been made public. Perhaps there is no evidence of collective or corporate negligence in this event. Such is the problem with employing actual events for philosophical examples, but, nevertheless, it is not difficult to recall other instances when negligence was uncovered and arguments subsequently made for holding someone other than the actual perpetrator(s) partially responsible for the deaths and damage. Events at Love Canal and in Bhopal again come to mind. In such eventualities, the question becomes how to identify those responsible for harm arising from their implementation of the technology directly relevant to the damage, that is, in our initial example, those involved in constructing the skyscraper in the first place.

Obviously, constructing a building such as the one brought down in Oklahoma City involves at least two types of collectively organized groups: private corporations and government agencies (the latter both as agencies involved as regulators of all such construction—such as OSHA and local inspection and zoning agencies, as well as those directly involved in the construction of a *federal* government building). The irony of this being a government building should not be lost when discussing the collective responsibility issues contained in the tragedy. In fact, this real-life example demonstrates, better than any hypothetical case ever could, the dangers of reification often present in ascriptions of collective responsibility that can lead—as in this case—to disaster. Early evidence from FBI investigation indicates that the jailed suspect in the bombing was acting from a motive of "general hostility toward the government," which he held responsible for his failed military career.[24] Generalizing the government's responsibility as the possession of a singular, *collective* entity made it possible for him to lash out at his nemesis without ever seeing the individuals who do, after all, make up the government and who, of course, occupied the federal building. Thus, he killed without remorse, taking refuge in the reification that portrayed his victims as merely a collectivity, not as individual people doing their jobs.

No serious (and sane) model of collective responsibility commits so palpable a fallacy of reification. Still, when looking at the collective responsibility at play in government agencies or corporations, it is difficult to avoid some level of reification when, especially in the case of corporations, the law itself treats them as "persons." Why is it that corporations are given such status? The usual response is to guarantee them the rights of property and commerce that corporations must exercise in order to achieve the economic goals of their members, goals also presumed to be beneficial for society. Clearly, with those rights come responsibilities, as is true for any individual's rights as well. When harm results as a consequence of corporate activity, the question is rarely whether corporations have responsibility for the harm or the risk of harm to others. A century of tort and liability law has already established that they do. The questions for philosophers and us as well are, first, why is a concept of collective responsibility necessary and, second, how should we construe responsibility so as to avoid the dangers of reification on the one hand and reductionism on the other?

The sizable body of literature on collective or corporate responsibility offers several, in some cases competing, justifications for the concept. I do not want to engage all of the arguments and so present what I consider to be a most relevant and persuasive reason for why we need—and, whether we know it or not, intuitively accept—the idea of collective responsibility. It is a reason that comes directly from the Oklahoma City bombing and from our general discussion of emergent risks. Because such risks, and whatever steps are necessary for alleviating them, tend to span generations, we need the notion of collective responsibility to protect the interests and even rights of future generations who will occupy the buildings we construct and, indeed, the entire environment we populate. There really is not a distinction to be made here between governments and corporations—both exist in the eyes of the law as collective persons with responsibilities to their members. Obviously, governmental institutions and their responsibilities for the public welfare are generally assumed to continue beyond the life or service of any present occupants of government positions. Indeed, it is a major reason legitimating the very idea of government that it guarantees such continuity. But it is the institutions that continue, in more or less recognizable form, generation after generation. Congress's responsibility for legislating, for instance, belongs continuously to the institution regardless of its occupants at any given time.

In similar fashion, the responsibilities that corporations should (and often do) assume for protecting the environment or to guarantee the future safety of their buildings (to take our examples) are predicated on an essential contract of sorts between persons living today and those in the future whose interests need protecting. Within liberal theory, however, it is difficult to assume that an individual person living today can be held to any notion of "contractual" accountability to someone not yet born.[25] In such instances, as Peter French argues, it becomes necessary to realize that "there are persons that exist now that can have actual relations with the generations of the future. They are corporations or corporation-like entities."[26]

Not all theories of collective responsibility rely solely on this attribution of moral obligation to future generations. But all of them, as I contend with French, must imply such a responsibility as part of whatever their argument for holding corporate entities responsible for their actions. And, surely, this is a primary reason why corporations are viewed in law as individual persons with legal standing. As such, they are accountable at law to persons outside the organization, either at present or in the future. Like all other citizens, those corporate entities are entitled to run risks in order the achieve their own ends; thus, the importance of collective and corporate notions of responsibility for our ability to cope with risk issues is clear. Emergent risks of the technological type we have been discussing are characterized by an often indeterminate starting point and in many cases a very long life (thousands of years in the case of nuclear decay of radioactive wastes). Therefore, our need for responsible corporations and government agencies is pressing indeed.

Understanding why a notion of corporate and collective responsibility is needed does not automatically solve all the problems of how to conceptualize it. In general, most philosophers and ethicists view collective responsibility in one of two ways. First, following Peter French, many construe corporate responsibility in terms of the collectivity taken as a whole and treated as a "moral person," much as how the law treats corporations as legal persons. In French's view, corporations are more than just organized groups of individual people; they "have a metaphysical-logical identity that does not reduce to a mere sum of human-being members."[27] Seeing corporations this way allows treatment of them as full-fledged members of a moral community shared with individual people and thus bestows upon them "whatever privileges, rights and duties as are, in the normal course of affairs, accorded to all members of the moral community."[28]

French's approach carries the benefit of treating collective responsibility in a way recognizable to those steeped in the liberal individualist tradition. Although corporations are not *reduced* to the individuals within them, they are treated as individuals themselves, and thus the traditional language of rights, contract, and responsibility can be employed in our dealings with them.[29] However, as Larry May points out, the problem with this approach lies in its lack of distinction between types and degrees of responsibility for corporate behavior when compared to human behavior. As collective entities, corporations are not the same as human beings from the standpoint of moral responsibility, for the actions of corporations are contingent upon the actions of human beings, although the reverse is not necessarily true. Thus, corporate actions are always joint actions, whereas human actions can be solitary. In May's opinion this difference causes the analogy between corporate and individual responsibility to break down.[30]

May's insight is often important in the legal determination of corporate liability for the imposition of risk or harm. Because a corporate member's culpable behavior might be construed (most likely by the corporation itself and its lawyers) simply as a private individual not acting in his capacity as a member of the corporation, the corporation might succeed in avoiding legal penalties or restitution for the harm done. In the Bhopal disaster, for example, Union Carbide attempted, with some success, to paint itself as a collective victim of one or a few of its disgruntled members. In light of this commonly recurring legal gambit, Dennis Thompson offers a different model for construing responsibility within collectivities.

For Thompson, the problem with French's approach to collective responsibility is that when applied to public bureaucracies it effectively undermines the democratic goal of accountability of public officials and, ultimately, the whole idea of moral responsibility as well. Viewing public agencies in the same way that French construes private corporations would, as Thompson sees it, lead to the "problem of many hands," which he describes as having three stages:

> [M]any political outcomes are the product of the actions of many different people whose individual contributions may not be identifiable at all, and certainly cannot be distinguished significantly from other people's contributions. The second step is the claim that no one individual, therefore, can be morally blamed for these outcomes. At the final stage of the argument, its proponents reach two seemingly contradictory conclusions: one stating that every individual associated with the collectivity should be charged with moral responsibility, the other holding that only the collectivity can be so charged.[31]

These two conclusions essentially deny the whole idea of responsibility, Thompson claims, because neither ascribes responsibility on the basis of who caused or, even was in a "specific and distinct" connection with the harm.

Thompson's approach seems particularly sensitive to the "emergent" phenomena within social groups, where individual contributions to harm (or, for my purposes, risk) are not easy to identify. In his important response to this difficulty in assigning responsibility, he manages to avoid both fallacies of reification and reductionism in his solution. He argues that we can isolate cases of individual responsibility, despite the emergent problem of many hands, but only by interpreting the requirement of causation very weakly. Thus, an individual official can be held morally responsible and therefore blameworthy if the harmful "outcome would not have happened but for one's act of omission. To say that a person is a cause merely connects his or her action with the outcome—along with the action of many other hands and the influence of many other forces."[32] Furthermore, to arrive at a stronger causal claim and therefore to hold any one person especially responsible, we must "invoke other moral and political considerations," prominently, the relative desirability of the outcome.

Thompson does not deny what French calls the "metaphysical-logical identity" of an organized collectivity and realizes that within the organization an individual officer acts in ways prescribed by the role he performs. Yet officers are still individuals and therefore accountable (in the case of public officials to citizens rather than consumers) on moral grounds for actions that occur at least partly because of their actions or omissions. The weakness of the ascription of causality places perhaps less stringent moral requirements on the member of the organization acting as such than on individuals acting purely on their own initiative. Still, there is responsibility being assigned here, and it is not being relegated to the collectivity as a whole.

Larry May adds to Thompson's model by referring to the responsibility that emerges from many hands as "shared responsibility," by which he means responsibility that "distributes to each member of the group."[33] Unlike collective responsibility, shared responsibility presumes that individuals within the group can be identified as having contributed to a result and therefore are responsible jointly to bear the burden of liability. Unlike individual responsibility, shared responsibility acknowledges that group membership changes things:

> Group membership changes a person's normal duties and responsibilities. People join groups because the groups enable people to do things they could not otherwise do alone. By joining groups, individuals attain increased ability as a benefit, but there is also a corresponding cost: members have a special duty to prevent harms caused by their fellow group members, where such preventative acts do not risk significant harm to group members.[34]

Together, May's and Thompson's versions of group responsibility make it possible to recognize the individual contributions to collective action and, therefore, to collective responsibility. In the process, they also hold out the possibility of our being able to assign differential shares of responsibility for corporate harm to different members of the corporation.[35] This is an added bonus, but the primary benefit of seeing through the collectivity to the responsible individual members within it is not a reductionistic one. Rather, viewing collective responsibility as shared allows us to expand our assignment of responsibility to include even individuals outside the organization whose desires or needs gave rise to the collectivities themselves. As citizens, this means that we must all assume some (albeit perhaps weak) causal efficacy, and therefore some measure of responsibility as well, for governmental actions. As consumers, this means that we also share some of the responsibility for the risk arising from technologies we demand and employ in pursuing our preferred way of life. In other words, the risks of skyscraper technology are shared by all of us who depend on and enjoy the lifestyle that presumes the technology being utilized. It is difficult to see who among us might be exempt from this responsibility. Precisely this realization—that all citizens are complicit in the risks of skyscrapers—makes the decision of citizens of San Francisco and Seattle to use the referendum decision process especially "responsible." All citizens must participate in the decision, because all of them, to one degree or another, participate in the generation of the risk.

Risk and Shared Vulnerability

The emergent character of modern risks makes responsibility for them rely only on the weak causal efficacy identified by Thompson's and May's views. As a result, responsibility for risk can be distributed in both a wider fashion and in differential measure across individuals and corporate entities within society as a whole. What this means, briefly, is that for many risks all citizens bear some responsibility but not to the exclusion of additional assignments of responsibility to specific individuals or corporations. Thus, differential levels of accountability, blame, or reparation can be demanded from different citizens (human and "legal") in society, although everyone can be expected to shoulder some responsibility for coping with the risks of modern life. How much blame or responsibility any individual or collective body should shoulder is partly a political matter, as well as a moral one, and therefore cannot can be determined beforehand or by application of general rules. Responsibility is case-specific and is, of course, subject to the democratic processes in place in society. Those processes are themselves greatly affected by risk, and, as we will see in chapter six, need much revision.

It might be objected that such a view of responsibility essentially divorces responsibility from causation altogether and in the process blames everybody for harm that further investigation would locate as emanating from a singular and identifiable causal agent. Thus, the objection might conclude, we all end up paying for what is really the fault of individual bombers (in our example in Oklahoma City) or otherwise readily identifiable agents of risk such as Union Carbide; Dow Chemical; the Bureau of Alcohol, Tobacco, and Firearms; Genentech; or J. Robert Oppenheimer—just a few entries on what could be an endless list of guilty perpetrators of modern harm and risk.

My response to this objection is twofold. First, it is important to understand that to say a risk is emergent already acknowledges the almost countless number of "many hands" that had a part in its creation. In other words, emergent risks are often visible, as we saw in chapter one, only at the level of culture and society as a whole. At least, they are often not acknowledged as risks until they reach a level of saturation that makes even the most minimally informed citizens aware of their presence. At this point, individual agents or causes may no longer even be relevant, as the risky practice is now institutionalized at the level of culture. Thus, for example, it makes little sense today to hold Henry Ford responsible for air pollution or J. Robert Oppenheimer (or Albert Einstein) for radioactive wastes. We might still invoke special levels of responsibility on Ford Motor Company or modern utility companies, but this does not diminish the responsibility of all drivers to keep their cars running as cleanly as possible or of some citizens to allow a waste facility in their town.[36] To insist on such a large number of causal agents may seem disconcerting because there is really no one left to sue, but *not* to do so really does divorce responsibility completely from cause. The truth of the matter is that lifestyle choices that include driving cars and using electricity are important causes of risk; thus, responsibility—like causation—must be shared.

If we do not accept that all individuals share responsibility for emergent risks, our only recourse is to blame technology itself as somehow autonomous and beyond the reach of our arm of enforcement. As we saw in chapter three, this bit of reification is fallacious. Emergent risks are not just the products of human endeavor, but collective endeavor that is thus not always—or usefully—reducible to the individual actions or persons contained within it. There *is* cause and therefore concomitant responsibility, but neither is entirely reducible to individual actions or persons.

The moral force of modern risk, due to its emergent character, is subtle, in that it brings us into each other's presence and organizes us in an unusually abstruse way. But it *does* organize us, so that in the face of modern risks, we are not simply a random collection of individuals who entail no responsibility. In the terms of a well-known example in moral philosophy, we are not analogous to a group of individuals on a beach watching a child drown while not doing anything to help.[37] The differences are two; first, our past and present choices as a group have put that child out there in the water and in jeopardy and, second, we are all that child. Emergent risks organize us into a recognizable group possessing causal efficacy for the risks that arise and within which individuals are each accountable for their share of any solution.[38]

As Virginia Held points out, the difference between organized groups (that can be held responsible for harm emanating from their actions or omissions) and random collections of persons, none of whom decides to become a "good Samaritan," often is that the former demonstrates the feature of established authority capable of organizing action.[39] Thus, in the case of blameworthy omissions, culpability lies in the failure to organize sufficiently so that authoritative decisions can actually be reached. As we saw in chapter three, when it comes to risk, authority in liberal societies is often not very "established;" thus, liberalism is blameworthy for its failure to confront risks. Yet the consequences of modern risks—their "authority of ends"—do tend to promote collective action on the part of groups of individuals confronting the risks of modern technology, as the widespread benevolent response to the disaster in Oklahoma City has once again demonstrated.

This organized response to the consequences of risk, so often present in the catastrophes of our day, constitutes the second reason why we should not be persuaded by the objection that shared responsibility essentially renders causation irrelevant as a factor in determining blame. In short, the countless stories of distant citizens responding with gifts and assistance in the face of disaster is an acknowledgment of responsibility that carries a sense of sharing in the causation of the event itself. Television technology (or better, modern communications generally) enhances this sense of communal responsibility, but it is responsibility born out of a special sense of cause. We respond with sympathy and help for disaster victims not solely out of relief that the randomness of events has somehow spared us and not them. Rather, we know that we all rightly participate in the event somehow—some of us by being there, some vicariously by watching on TV—because our shared technological knowledge (and our share in its benefits) ultimately (though weakly) causes such events. We are all participants in a technological culture that has produced and that shares the knowledge to make towering buildings, as well as towering infernos, possible. Thus, we all participate in the cause and in the responsibility. In the face of tragedy, we acknowledge our joint, communal causation as well as the authority of the consequences themselves that organize us into responsive—and responsible—action.

Perhaps this argument seems too much of a stretch. Do we really want to say that because watching television involves us in a sharing of technological know-how, we are by that fact alone responsible to help all those we witness suffering the tragic unfolding of modern risk? Although this might provide a good argument to discourage excessive television viewing, it may not seem to argue a good case for distributing moral responsibility. What of those who do not have or watch television—are they somehow exempt from the "shared responsibility" of coming to the aid of disaster victims? Does buying a television then invoke a special contract with every other person to come to their rescue if we see them suffering on the six o'clock news?

Objections like these betray a moral incredulity born of an excess of liberal individualism, made outdated by modern technological life. We need to realize that the complexity of modern life requires a more sophisticated notion of cooperative action and communal responsibility. Life today is so full of risks and potential harms about which issues of responsibility pertain, that unless we move beyond liberal notions of individual responsibility, we will, argues Annette Baier, "bog down on endless disputes about just who should get what portion of the blame for past evils, about just who deserves what compensatory advantages now."[40] As Baier warns, the dangers are very real if we hold on to our outdated notions of individual and corporate responsibility that cannot see beyond reductionist notions to truly collective, shared, emergent responsibility:

> If we insist on clinging to the idea that moral responsibility must divide without remainder into the bit that is mine and not yours, and the other bits that belong exclusively to specific individuals, then not only will we limit the sort of shared action we engage in but we will drastically limit our ability to reform our inherited schemes of cooperation for the better. . . . The Kantian form of individualism that has been and still is being appropriated in the American tradition has two fatal and connected flaws. It has no account of really shared responsibility but only pooled or passed-along individual autonomy and responsibility.[41]

Baier's warnings constitute important support from within mainstream moral philosophy for the alterations I am suggesting to our notion of responsibility. Nevertheless, they do not entirely convince that knowledge of suffering by itself is a sufficient warrant to assign responsibility to individuals or to a community as a whole for its alleviation. If knowledge alone is insufficient, then the technological expansion of that knowledge via communication technologies also does not increase the extent or range of responsibility to include all who are aware of the suffering. Our modern technological ability to bring the knowledge of suffering to a larger audience may then, at most, be a contributing factor in the expansion of responsibility. Something else is still needed to make a convincing case for my being responsible to help someone suffering thousands of miles away simply because I have knowledge of his or her suffering.

Robert E. Goodin confronts this objection head-on by exploring why moral theory has traditionally assumed a more stringent responsibility to assist those closest to us than those at a distance. Goodin is seeking a method of grounding responsibility that will expand personal moral responsibility to include obligations to others far removed from us in a variety of senses—in terms of miles, time, even position on the evolutionary ladder. Eventually he provides arguments requiring personal responsibility for the welfare state, for foreign nationals in need of aid, for future generations, and for stringent environmental protectionism. The responsibilities arise from the same basis, Goodin contends, as those traditionally accepted by most moral theorists to family, friends, and contractual partners.

Goodin collapses all distinctions between these supposedly different responsibilities into a "unified, coherent ethical principle" that underwrites all responsibilities one has to others at whatever distance. That principle is resoundingly similar to the whole idea of risk: it is what he terms "vulnerability . . . a matter of being under the threat of harm."[42] It is Goodin's contention that all moral responsibility arises from the same source, namely, that others are vulnerable to our decisions and actions—what we do puts them at risk and we are responsible therefore for the consequences to them. He writes:

> What is crucial, in my view, is that others are depending on us. They are particularly vulnerable to our actions and choices. That, I argue, is the true source of all the standard special responsibilities that we so readily acknowledge. The same considerations of vulnerability that make our obligations to our families, friends, clients, and compatriots especially strong can also give rise to similar responsibilities toward a much larger group of people who stand in none of the standard special relationships to us.[43]

Goodin's argument is one particularly well-suited to the question of moral responsibility in our age of emergent risk. As he recognizes, this is a moral theory particularly sensitive to the authority that consequences (or ends) exercise in the lives of people aware of the interconnectedness of modern life. Some of these interconnections have not been assumed voluntarily or contractually by individuals living in modern technological society. Rather, these connections have emerged and have been recognized gradually by means of, among other things, the effects of modern communications technologies such as television. Thus, by watching the events in Oklahoma City we sense a responsibility to help that we may not have consciously bargained for but nevertheless acknowledge. This acknowledgment is, as Baier contends, a denial of the Enlightenment liberal grounding of responsibility purely on voluntary acceptance implicit, or so it would seem, only in the special obligations to spouses, children, or business partners.[44]

As Goodin himself recognizes, in addition to the Enlightenment liberal version of voluntarist or contractual obligation, he is attacking other approaches to moral responsibility. He spurns traditionalists insisting that different responsibilities are based on entirely different and unique sources of moral obligation, as well as the Kantian view that rejects the obvious consequentialism of the vulnerability model. Yet Goodin successfully shows not only the superiority of the principle of vulnerability in terms of usefulness but also how it can yet defend special obligations sought by the traditionalist and Kantian schools.

It is not my intention to recount Goodin's argument in its entirety or its considerable subtlety. But using vulnerability as a measure of to whom (and, for Goodin, to what) we are accountable for our actions goes far in explaining how emergent risks alter our understanding of responsibility in ways unanticipated by liberal moral theory. Utilitarian liberalism is most amenable to Goodin's approach, yet when compared to the general idea of maximizing happiness, vulnerability (or, in my usage, susceptibility to risk) does a better job of concretely directing moral behavior and also better anticipates issues of whom we should take into account when contemplating (or compensating for) risky behavior.

Modern technologies make us vulnerable to each other in a variety of ways unique to our time. They pack us more closely together into cities and tall buildings, and they bring us into each other's lives by means of television wires and computer cables. In so doing, they display our mutual dependency in all its subtlety and its inescapability. As we share our vulnerabilities or risks, we must behave differently in terms of our responsibility for the harm done or threatened to others by our actions. We must, in other words, share our responsibilities as well. It is not Goodin's contention—or my own—to insist that the fact of emergent risks and vulnerabilities means that we must now treat every person or even every species as we would our spouse or our children. Our familial responsibilities can retain their specialness even as we acknowledge our obligation to persons further removed from us but still vulnerable to our choices and our behavior. We can, in other words, still insist on different levels of vulnerability that give rise to different degrees of responsibility, levels and degrees that are, at least in principle, discernible and distinguishable in terms of degree of potential harm that one's action causes.

Goodin's theory, and my special application of it, does not deny completely the causal basis of responsibility. Yet, like the positions of Thompson and May, it accepts that vulnerability can admit degrees of causal efficacy that distinguish variable levels of obligation. For instance, due to the wealth of corporations' pooled resources and the range of their harmful effects, they bear special—and higher—responsibility for the consequences of their actions. Furthermore, because individually we are capable of causing the most harm to those closest to us, our responsibility to them is still special under the vulnerability model. But this approach to cause and responsibility acknowledges too that even our most passive and amorphous actions, such as those involving lifestyle choice, also have consequences that bring other persons (and creatures) into our orbit of vulnerability. Thus, we must acknowledge responsibility to them as well. They are distant, perhaps, but closer than we once thought. What makes them so? The increased knowledge that modern science delivers and also the web of emergent risk that modern technologies construct, televise, communicate.

Conclusion: Risk, Luck, and Communal Responsibility

In recent work on moral theory, Marion Smiley argues, in opposition to Feinberg, that in decisions about responsibility, questions of both cause and blame rarely involve judgments "as purely factual as we often assume." Instead, both "rest on a variety of social and political considerations which we either make ourselves or inherit from others in the form of social expectations."[45] Smiley criticizes modern moral theory that, in her view, begins with a rejection of both Christian and classical understandings of responsibility grounded on relationships between individuals and an external blamer, either God or the community. Instead, modern theory "denies the relevance of such a relationship and locates the source of moral blameworthiness in an individual's own free will," thereby rendering the modern view of responsibility essentially "sinfulness minus God."[46] Such a view overemphasizes causation as the critical issue in assignments of responsibility, asserts Smiley, and underplays the social context, or the moral community within which responsibility matters.

By rejecting what she sees as the modern view, Smiley in large part denies, on somewhat different grounds from mine, that causal efficacy is an important part of any socially or politically useful theory of responsibility. She opts instead for a concept of social blaming that will allow ascriptions of responsibility when socially useful in order to enforce certain necessary behaviors without assigning guilt as some kind of stigma or "black spot on one's soul."[47] A combination of utilitarianism and pragmatism, Smiley's approach has much to recommend it in areas of emergent risk, given the difficulty of determining cause in such cases. Smiley goes further than either Thompson or May in rejecting causal issues as relevant, and we need not travel all the way with her to recognize that, in some issues of risk, her approach is quite fruitful, particularly when combined with Goodin's vulnerability model, a possibility she also embraces.[48]

For instance, in siting disputes involving the location of hazardous waste dumps, it is often the case that geological factors make site selection somewhat easy, whereas political issues make the decision almost impossible. In such a common scenario taken from today's newspapers, the issue of responsibility is irresolvable if approached from the standpoint of modern moral theory. Citizens living in proximity to the site cannot be said to have *caused* the problem of hazardous waste, at least to any greater degree than other citizens. Thus, why should they shoulder the responsibility for alleviating the dangers that wastes portend? To force them to do so seems a violation of their rights to be treated as equals, especially in light of the fact that the risks associated with hazardous wastes are emergent in nature and thus implicate us all as generators of both the wastes and the risks they carry.

Many theories of responsibility that highlight issues of causation rely on such a concept of equality largely because, if they do not, elements of luck enter into the equation that seem to render the whole situation outside the realm of morality.[49] If someone is unfortunate enough to have built a home on (it later turns out) a geologically perfect site for a dump, is that reason enough, such theories ask, to force him or her to accept a social responsibility to allow the site to be developed? For most modern moral theorists the answer would be no, because the individual citizen did not cause the problem. For Goodin, Smiley, and me, other factors might at least in some cases argue that, yes, such a citizen has a responsibility to allow construction.

According to the vulnerability model of responsibility, the obligation to allow the site is directly related to the degree of risk of potential harm to the rest of society if the facility is not built. It is true that those living nearby are unlucky, but this in itself does not absolve them of their responsibility to the community. Their responsibility is also a matter of the vulnerability that might ensue if they do not allow the construction; it has little to do with whether they *caused* the risk. Of course, given our discussion of the nature of emergent risk, we know that in some way we *all* cause the risks that make modern technological life both rich beyond the dreams of our forebears and inexplicably threatening as well. Still, this is an extremely weak sense of cause, because it is incapable of singling out particular persons who are causal agents. It seems an interpretation of causality hardly strong enough to accuse those citizens of irresponsibility who respond to the social need with the NIMBY declaration.

Those who refuse to accept their responsibility in such cases are blameworthy, I contend, but not because they caused the problems associated with hazardous waste production and disposal to any greater extent than their fellow citizens. But, in refusing construction, they shirk a responsibility that is theirs perhaps through unfortunate happenstance, but nevertheless theirs. To say they have responsibility is not to say they are in any way guilty or blameworthy of the initial offense of creating hazardous waste—only that they are now accountable for the risk that others must live with if they successfully block construction of the disposal site. Responsibility and blame are matters between those citizens and the rest of society—between those who are to blame and their external blamers—not matters concerning moral facts about the individuals themselves.[50] Responsibility is to the community and invokes the community's vulnerability, as well as its moral authority. This also reaffirms the political nature of such moral issues of responsibility. In the real world of risk, we need not only make an argument that we are all responsible or that some of us are more responsible than we thought. We need actually to decide the real question: whose backyard? Answering this question brings moral issues into the everyday political life of a community. As we will see in the next chapter, risk thus affects the political discourse of modern democracies, both in terms of how decisions should be made and of who should make them.

Of course, in this example, some citizens' acceptance of their responsibility to live next door to a waste dump now makes *them* vulnerable, especially so compared with the rest of society. Thus, the community must in its turn recognize a special responsibility toward them. This is also implied in Goodin's approach (though he does not acknowledge it) and in any program of social compensation, whether involving those who risk all in battle for their country or who risk their children's health for the sake of society's sanitation or energy needs. As such, compensation must be forthcoming as a matter of social policy rather than through court decision in civil suits under tort law.[51] Society *as a whole* now owes a debt of gratitude to those citizens made vulnerable by its need for disposal sites and by their acceptance of their special (though unlucky) responsibility to live near the site.

There is simply no escaping the conclusion that modern technology and its attendant risks entwine all citizens in a web of responsibility that is at its center both political and moral. The bad luck of siting disputes is merely the current manifestation of the twentieth-century truth that "whose bad luck it is—who should bear the cost—is the fundamental question of political philosophy."[52] What makes it a moral question as well is not only the issue of one's debt to the moral community made vulnerable by waste (or other risks) but also of the community's debt to those willing to assume greater vulnerability in order to diminish that of society. These are essential questions of politics in every time and place because they require the exercise of power and the weighing of interests.

As moral issues these questions engage all citizens in a participatory system of ethics that is both consequentialist and deontological. Responsibility for risk focuses our attention irresistibly on consequences but also erects new rules requiring us to expand our circle of specially felt obligations to include those about whom we know little except that they are vulnerable. In other words, modern technologies and their emergent risks have expanded the purview of our moral awareness as they have broadened our ability to affect our world. Consequentially, they allow others to participate in our lives in ways both deleterious and beneficial, as we do in theirs. The choice so to participate has already been made—it is (was) part of the emergent decision a society renders in embarking on technological development. What will be the moral effects of participation in each other's lives is a question of politics that, in a democracy, remains always in the process of being answered.

6

Democratic Politics and Participatory Risks

Throughout the foregoing chapters I have presented two arguments concerning the nature and impact of modern technological risks. First, I have suggested that if we are ever successfully to make policy for them, we need to realize the changed nature of these risks. No longer solely recognizable as the product of individual choices, modern risks should be viewed as emergent phenomena that resist reduction for explanatory or policy-making purposes to the individual actors, decisions, and environmental responses that together produce them. Second, I have recounted how, in the presence of such risks, political discourse has been transformed in a variety of ways, as citizens, experts, and politicians grope for policy solutions that commingle moral, technical, and political considerations. This discursive transformation is due to the alteration of the relevant terms of public speech brought about by the politics of risk. Within risk politics, core political concepts—including risk itself—have become matters of contention, disagreement, and, in some cases, of redefinition. As we have dealt with several of these terms of liberal political discourse—politics, consent, authority, rights, and responsibility—we have learned how disputes over risk policy have become battles over what these concepts are to mean in contemporary politics. Thus, the politics of risk has produced conceptual disputes over the definitions of risk and of some of the most basic concepts of liberal politics.

Political decision making within such a context of conceptual change is always difficult, because real changes in politics are usually also in the offing. Much of the history of domestic politics in the United States, in fact, can be written in terms of the carnage and subsequent reconstruction engendered by such conceptual disputes. At its founding, this country started a revolution over the meaning of political independence and individual liberty; that conflict's successful conclusion changed those terms forever. The Civil War began a hundred years of grappling with the meaning of liberal equality, during which time the nation struggled first with how ideas of equal liberty affected slaves, blacks, and women and later with the more amorphous concept of "equal protection" to define a new politics of social welfare and collective responsibility. The 1960s brought its own sometimes violent conceptual disputes over justice and the boundaries of individual duty.

In important ways, the conceptual struggles marking significant eras in US history and politics are still being resolved today, and our recounting of how modern risk has invoked some of the same issues and concepts only reaffirms the truism that democracies never really finish defining what it is for which they stand. Predictably, then, as we have seen in many areas of technological risk policy, decisions about risk have simply been deferred until terminological disputes of both technical and political natures can be resolved. These disputes, as Terence Ball and J. G. A. Pocock point out, are "apt to be hard-fought, in a most ungentlemanly way, with almost any weapon that comes to hand."[1] Such battles have attended risk decisions precisely because risk itself can no longer be seen as the result of exclusively individual decisions and actions. I have therefore recounted how liberal political concepts grounded in an individualistic ontology are of little use in defining answers for a political issue manifesting new collectivist and emergent characteristics. Risk politics highlights questions of consent, authority, rights, and responsibility but demands new conceptualizations of them.

Political theorists may think it heartening that within risk politics the conceptual struggles surrounding liberalism's foundational ideas have permeated the body politic and elevated its internal discourse to a more philosophical level. Political theory might then itself be revitalized by risk politics, since political theorists offer an ability to mediate and clarify disputes about the concepts behind political events and behavior.[2] What rights are to mean, how far authority extends and who should wield it, upon what basis are both obligation and responsibility built—these are questions embedded within risk politics and batted about not only in arcane academic journals but, as we have seen, in courtrooms, in zoning board meeting halls, on the Internet, over the talk radio airwaves. The discussion is indeed not always "gentlemanly," but its lack of restraint nevertheless manifests one of democracy's most endearing and essential features: its openness to new ideas and to challenging the meaning of its oldest, most cherished beliefs.

Yet, in another way, it is ironic that this raucous discussion centers on the concept of risk. Liberalism has traditionally construed the domain of risk to be that of the individual citizen privately deciding a future course, weighing the costs and benefits in good enlightenment-rational fashion, being free to choose (or not) the way that leads through risk toward private reward. Because we have seen that modern risks present an emergent and eminently social reality rather than an isolated and privately chosen one, today the very identity of the liberal citizen is challenged when confronting risks. Once identifiable as a solitary, even heroic (if mythic) figure heading into a risky and uncertain frontier, protected only by the certainties of rights, individual liberty, and the accountability of whatever authority encountered along the way, today the liberal citizen ironically has been remade by risk. Now he or she is beset and surrounded by the nonindividuated and therefore almost unrecognizable risks that exert power over choices, set limits to exploratory liberty, and threaten to entangle him or her within a web of attachments and responsibilities only dimly recognizable as the product of her own actions. Within such a context of subtle power and collective impingement, it is no wonder that what emerges as discourse about the new realities of liberal freedom, consent, rights, and the conceptual rest appears neither philosophical nor elevated but looks instead like mere shouting and the venting of liberal frustration within a not very civil realm of civic participation.

In this chapter I want to explore what the conceptual changes wrought by technological risk portend for the identification and practice of liberal politics. From what has already been said, it is clear that much stands in need of redefinition if liberalism is to learn to cope with the fact of modern risk. Since by coping I mean both surviving and becoming able to make policy for the regulation of risk, liberalism needs to adapt to new procedures for decision making. Risks are emergent; decisions must become so also. That will mean new roles for citizens as well as for political leaders and technical experts working in concert to address the multiple policy facets of most risks. Thus, in addition to exploring the emerging concept of identity in technological societies, we must also explore the changes in the meanings of citizenship and democracy, changes made requisite by modern life's risks.

In what follows I presume it is uncontroversial that, despite the challenge risks pose to democratic decision making, we persist in valuing democracy as the mode of decision. Many of the risk policy dilemmas we have explored in this book might be easily resolvable simply by suspending democratic procedures or guarantees and by delivering the power to resolve disputes into the hands of a benevolent despot. Democracy is itself in jeopardy because of modern technological risks, in part because the complex character of those risks makes it tempting simply to throw up our hands and allow someone willing to decide to assume the power necessary to do so. Democracy requires perseverance, however, in always resisting the appeal of the Grand Inquisitor, whether clad in clerical, royal, or scientific garb. In the case of risk, perseverance will take the form of increased citizen participation in the decisions required of modern societies facing technological hazards. Since the risks affect us all, truly efficacious decisions at every level—political, moral, psychological—must engage us all as well.

The following discussion delineates the democratic response to risk along three dimensions. In the first section, I explore how modern risks feed into the current growth of what is usually called "identity politics" within liberal democracies. Many risks define groups of individuals in terms of the risks they share, thereby presenting them as identity groups with demands and interests potentially at odds with society as a whole. In so doing, modern risks also illuminate liberalism's difficulty with the entire concept of group identity, as well as its disabilities in making decisions about group risks.

In the second section of the chapter, I discuss how risk's effects on identity require that liberalism uncover a new notion of citizenship that I call "emergent citizenship." Because risk's effects on identity and on politics can only be identified at the emergent, group level, citizenship itself needs to become less reliant on individualistic concepts such as rights and opportunities and embrace a more communal sense of duties and participation.

Finally, the last section explains what technological risks portend for modern democratic political practices. Risk places identity groups at odds with each other in their struggle publicly to present and protect not only interests, but identity itself.

Democracy, then, should be defined no longer as simply the pluralistic competition of interests but as an "agonistic" realm where disagreements can be negotiated only if all citizens accept a stronger claim of community than that liberalism traditionally has embraced. Here the emphasis is not only on creating more new avenues of democratic participation for citizens in the policy process for risk but more generally on reconfiguring the "civic realm" within which citizens encounter each other both as private and public persons.

Risk and Liberal Identity

We saw in chapter two how within liberalism the tension between the values of autonomy and reciprocity resulted in a communitarian conundrum denying to the liberal citizen an identification with his or her community that is truly constitutive of individual identity. We also observed how liberalism's reliance on tacit consent as the basis of political organization allows for the sharing of interests but not of selves, of common purposes but not a common conception of the good. More complete sharing—that would be definitive of a fuller sense of community—demands expressed consent, but such consent presumes either a loss of autonomy unacceptable within liberal theory or else institutions legitimated by a level of ongoing participation truly expressive of conscious consent. These institutions are not widespread within most liberal societies, and, indeed, political trends seem aimed away from them. Yet finding solutions for modern risks demands that individuals identify in more than just an instrumental way with their communities. If they cannot, then liberalism has a serious policy problem in coping with risk that emanates from its failing to provide a strong enough sense of communal identity.

If, in other words, solutions to risks such as those emanating from the storage of hazardous wastes (to take an inescapable example) are ever to be arrived at, the notion of communal good must at some point emerge as an ascendant social interest outweighing the private interests of individual citizens. At the point where social interests gain ascendancy within the minds of individual citizens, thus making effective disposal policy possible, their sense of self-identity should already have been transformed in the direction of the larger community.[3] Liberal politics has yet to achieve this identity transformation, and therein lies its dilemma in making risk policy.

I argue in this chapter that the impact of modern risks is slowly bringing an awareness of the outdatedness of the liberal notions of community and individual identity. Risks are forcing a conceptual change—a sea change really, for what they transform are not only interests but senses of self and others, that is, perceptions of identity. Perhaps solutions to modern risks might yet be found without compromising the core liberal values of autonomy and reciprocity. Those values must be resettled, however, into a new notion of citizenship and within a new democratic civic realm where participation can redefine consent to include an *expressed* commitment to the welfare of the community as a whole. The subsequent sections of this chapter elaborate these new views of citizenship and democratic participation; first, we focus on the conceptual evolution of individual and communal identities as manifested in the response to modern risks.

Any consideration of modern technological risks draws our attention to the degree to which important features of individual identity rely upon actions and decisions outside the control of individuals. Also, disputes stimulated by the distribution of risks (such as siting controversies) show that questions of identity always evoke power and are thus essentially political (and essential political) issues of our time. As emergent phenomena, then, risks in their etiology as well as in their effects blur distinctions between individual and community, between self and others. Thus, they are at the leading edge of the politics of identity in all its forms, whether of locality, gender, nationality, ethnicity, or minority group.[4]

Before exploring several examples in which specific risks give rise to identity politics, we should briefly recap why the liberal notion of identity is clearly insufficient for the contemporary purpose of coming to grips with the politics of risk. Within the constrictive limits set by its values of autonomy and reciprocity, liberalism offers both a too abstract definition of individual identity and a too shallow definition of communal identity.

Individuals are said to be worthy of autonomy, if one follows Locke and Kant, because of generalizable features possessed by all human beings—reason, perfectibility, understanding of the natural law, moral intuitions, whatever particular form of the liberal argument is being used. But this is obviously too abstract since, from the individual's standpoint, it says little about what individuates him or her from others. For instance, to be authenticated both to myself and others, my identity needs differentiation from others' in terms of what I *am*, not, to follow Connolly, "what I choose, want, or consent to." But the features underlying autonomy are only abstract underpinnings of what makes it possible for me to choose, want, or consent—that is, reason, perfectibility, and the rest. If, for example, it is reason alone that leads me to my own choices, but reason is, as for Kant, universal, then how or why would my choices be any different from anyone else's? That is, why would I be different, much less unique? Here liberalism is silent about what it is about me, and me alone, that leads me to choose what it is I, and only I, choose. These choices invoke my "dense self from which choosing, wanting, and consenting proceed."[5]

What little liberalism has to say about Connolly's "dense self" has largely to do with liberal tolerance for potentialities—for how one uses and puts together the abstract attributes distributed to all individuals. Liberalism, in fact, makes no attempt to describe what any particular self might look or act like; to do so would threaten the autonomy of the individual in the free pursuit of self-definition. Freedom here is construed to be largely negative in nature, that is, consisting of the absence of interference from others on the grounds of liberal tolerance for diversity. Self-definition remains a social pursuit of course, notwithstanding the liberal emphasis upon negative liberty, as Rousseau most acutely among liberal thinkers realized. Thus, something of a concrete nature seems to be at work in the process. But here liberalism escapes excess abstraction only by being shallow. What individuals in the liberal society share in their journey of self-definition are interests only, not notions of the good. Goodness is outside the realm of political endeavor, beyond the grasp of collective choice. Liberalism thus explicitly avoids providing what Charles Taylor describes as an "essential link between identity and a kind of orientation. To know who you are is to be oriented in moral space, a space in which questions arise about what is good or bad, what is worth doing and what not, what has meaning and importance for you and what is trivial and unimportant."[6]

As we have seen in chapter two, liberalism's failure in providing a full measure of community is due to its shortcomings in providing what Taylor calls a moral orientation. These shortcomings result from deliberate choice of course, because liberalism consciously opts for toleration of multiple conceptions of the good as a necessary product of the protection of negative liberty. But toleration is often not enough to secure the identity of groups within liberal society, just as it also does not secure community. Identity usually requires more than toleration; liberalism shies away from providing it.

An example of liberalism's substitution of tolerance for full recognition of difference is provided by Anna Elisabetta Galeotti, in her exploration of the 1989 "headscarf" controversy in French public schools.[7] Three Muslim girls wore the traditional Islamic *chador* or headscarf to school in keeping with Islamic law requiring that women keep their heads covered in public. French school officials initially ordered them to remove their veils, but in the resulting public controversy retreated and allowed them to continue wearing them. However, Galeotti argues that the reversal, on the official grounds of religious tolerance in France, failed to provide the "public recognition of collective identity" that minority groups seek in liberal society. Instead, the decision "fails to differentiate meaningfully between wearing the Islamic veil and wearing a funny hat at school."[8] The right of religious freedom granted the girls the opportunity to wear the veil, but this right belongs to the individual in her private realm of conscience, within which one might just as proudly wear a funny hat as a matter of self-expression. In other words, the girls were allowed to wear the veil as a matter of private, religious freedom, not as a recognition of their public identity as Muslims in a Christian society. Liberal tolerance, therefore, subtly undermines the public and collective identity of such groups by defining their differences as simply individual (and possibly trivial) choices guaranteed by their private rights.

In short, liberalism's "abstract individual"[9] as the carrier of certain rights describes no *particular* person, and liberalism's definition of politics as the pursuit (collective or individual) of interest superficially defines no single but *all* possible communal entities. In its alternating abstractness and shallowness, liberalism tries mightily to remain tolerant and therefore neutral on the whole question of identity—that is to say, liberalism denies that identity (like goodness) is a political issue at all. A politics that seeks to provide a strong sense of personal identity shared by all citizens runs the risk of intolerance, a vice that liberalism in all its forms reviles as a the chief enemy of individual liberty. But as we saw in our discussion of authority, consequences in the real world often carry their own political efficacy, and nowhere is this more true than in the politics of risk. Within that politics, the issue of public identity plays a crucial role—and a very disruptive one for societies trying to sustain, in the names of tolerance and negative liberty, the identity-avoidance behavior of classical liberal theory.

So liberalism itself says little on the topic of identity, whereas risk politics speaks volumes. Within risk politics examples abound in which disputes over the consequences of risk or of risk policy decisions have profound impact on the self-identities of citizens involved. In other words, the risks we take define *who* we are and what we believe, not merely (as in liberal politics) *what* the interests are that we might have chosen. In a sense there is nothing surprising in this statement even in terms of how liberalism construes individual identity. If individuals are free to pursue their interests, then risking their necks climbing mountains or flying airplanes clearly may lead to a construction of self as "mountain climber" or "pilot." The difference within risk politics is that the interests are rarely so freely chosen, and, even if they are, the notion of identity that results may not be. Furthermore, the sense of self that emerges from a confrontation with modern risk is an identity neither completely of one's own making or even of one's own discrete self; the presence of others is vividly evident both in its construction and its resultant appearance. Risk gives rise to conceptions of self-identity both more collective and more public than those with which liberal politics is comfortable.

To bring these somewhat abstract musings back to earth, I offer a series of examples from contemporary risk politics beginning with what seems to be an example of liberal, interest-based political conflict. A town is chosen as a site for construction of a risky or unwanted installation, either public or private, and the usual dispute erupts over the risks involved to citizens, their children, the environment, etc. Clearly this seems describable within the language of interests, but in many cases the disputation proceeds instead along the lines of the identity of the town. Statements abound of the type, "We are a rural town; this project will change us," or "We didn't create the problem and this isn't our waste; dump it somewhere else," or "This is a town with a center where people meet and make their purchases at the shops along the green; build your superstore (or mall) in some other town." Statements like these concern the self-perceptions of citizens as members of a community. They are expressing a sense of collective identity jeopardized by an external threat.

As citizens' private and individual economic interests are surely involved for good or ill in projects such as housing developments, waste dumps, or malls, so too is their sense of who they are as a community and as individuals within it. The threat brings them together in a way their interests do not and grants them, for whatever fleeting amount of time, a sense of self and of collective identity in need of protection. Furthermore, the risk may bring to the surface feelings of exclusion felt by some members of the community who eventually pay a higher price (in terms, say, of proximity to the project) or gain less benefit, depending on how the dispute is resolved. They, then, perhaps for the first time, experience a sense of alienation from the collective identity of the rest of the town—they are the ones whose welfare is ignored (in their own opinion at least) in the collective decision to proceed with construction.

A second example highlights the more subtle and potentially unnerving effects on notions of identity produced by the risks of modern medical and biomedical technologies. As we discussed in chapter four, current medical technological capabilities raise questions about basic definitions of life itself; of what it means to be a mother in the genetic, biological, or social senses; and of the dispensability of fathers altogether in at least some of those senses. When our definition of risk is expanded by modern medical science to include the risk of knowing, how do we calculate the import for one's sense of self to realize that an essential part of one's medical identity is as "carrier" of genetic disease such as Huntington's? How much has AIDS and the various tests for the disease affected the identity of homosexuals, both in their own eyes and those of their community? And what of victims of the disease who are not gay? Their struggle to present themselves as they are without stigma is a battle against prejudice made no less difficult in many cases by the medical facts, or by their exclusion from the larger—gay—community leading the fight for public recognition of their common plight.

These examples demonstrate again how risks can build a sense of collective, or at least, shared identity based upon the shared confrontation of risk. Such an effect can be heightened by knowledge that certain diseases (or conditions or choices) are restricted to identifiable populations. Thus, genetically heritable diseases create, at least in the medical community's mind, a distinctive group identity of potential victims. Of course, when questions of money for medical treatment or research arise, the sense of shared identification provided by the disease often leads to collective political action as well. In these cases—and they are many—the political interest shared by members of a group manifesting the same medical susceptibility becomes grounded not merely on an identification with the same point of view, but rather with the same, shared fate. When that group shares other cultural, economic, or historic ties, for example, American African-Americans who also share a susceptibility to sickle cell anemia, the mixture of minority politics and medical consequences can be a powerful force building collective identity.

Sometimes technological risks expose other bases for self or group identity, based not on genetic endowment or medical susceptibility but on social class and economic position. Arguments for "ecological justice" often accompany siting disputes and raise issues as to why the dumps always seem to be built on the poor side of town and the strip malls and bypasses never seem to encroach upon the more fashionable neighborhoods.[10] The veracity of such claims may or may not be at issue, but as they are increasingly made, the potential for shared identity construction grows.

Finally, some risks raise issues of national identity beyond those involved in conflicts between groups of citizens seeking public recognition in their own right. Indeed, some modern technologies effectively place the entire concept of nationality or nationhood in jeopardy. For instance, modern communication technologies are no respectors of national borders, as television shows and their ubiquitous commercials are beamed around the globe. Exposure to the values embedded therein pose threats to some cultures who do not share Western values, threats to their identities as separate and singular cultures. The creeping homogenization of the world's cultures through telecommunications strikes many outside the West as a new form of imperialism, an insidious risk threatening the very idea of difference—and therefore of national identity—itself.[11]

Similarly, one of the less noted impacts of the risks of nuclear power generation and its attendant waste problems is its presumptiveness concerning the staying power of national policy decisions—and even of the nations that make them. If ever the United States successfully determines where and how to store its radwastes, for instance, it will be a decision meant to last thousands of years, until the last half-life of radioactive decomposition is accomplished. Such a decision clearly invokes issues of intergenerational justice but, even more, of what makes a nation and what it is about a nation that might possibly be said to continue into the distant future. Many policies relevant to technological development and risk involve the future and the welfare of future generations—including policies about energy usage, agricultural development employing new biotechnological capabilities, the harvesting of rain forests, the spread of fluorocarbons into the atmosphere, the littering of outer space with rocketry accoutrement. These issues encompass risks and impacts that do not stop at the water's edge—they merge the entire globe into one interconnected ecosystem that extends far beyond national borders construed both horizontally and vertically into outer space. Even if sense can be made of the identity of one nation's future "citizens" a thousand years hence, and of the impact levied on their interests or rights by today's risk policies, how can one separate their identity from that of the rest of whatever humanity may still be around? Risk today portends global consequences that, according to Michael Ignatieff, have already changed "our locus of belonging" and "our language of attachments."

> Already, however, we can just begin to feel our old attachments, our old citizenship, being emptied of its rationale. All the changes which impinge upon the politics of modern states are global in character: the market in which we trade, and in which our economic futures will be shaped, is global; the ecology in which we live and breathe is global. The political life of nation states is being emptied of relevance by the inconsequence and impotence of national sovereignties.[12]

National identity and its attendant concepts of sovereignty and authority are difficult enough to conceptualize from one generation to the next; technological capacities and risks that stretch our perspective to include an indefinite future and a vastly more interconnected world clearly burden those concepts even further. Connolly calls this effect of technological risk the "vertical pluralization" of identity, "where a plurality of political identities are formed below the state, through the state, and across states, [and this] compromises the state as the final site of collective political identity."[13]

Three sets of questions and conclusions emerge from the preceding examples concerning the impact of risks on conceptions of individual and group identity. First, the conflicts permeating and exemplifying risk politics are not merely those of competing interests, at least not of interests construed narrowly as private, economic wants. Rather, confronting risk brings into question basic attitudes toward self, others, the community, and toward the belief systems that underwrite each of those dispositions. Most risk controversies include a technical component and an economic component around which competing sides may gather. But, as we have seen repeatedly throughout this book, epistemological, ontological, and normative concerns also come into play at all levels of risk politics, challenging the technical expertise, the political authority, and the morality implicit in risk decisions. These are disputes of a deeper nature; they challenge, for James Bohman, the "basic framework of moral assumptions and political procedures" of liberal democratic systems; they are liberalism's "deep cultural conflicts."[14] They represent what John Rawls refers to as *the* challenge to political liberalism: "Now the serious problem is this. A modern democratic society is characterized not simply by a pluralism of comprehensive religious, philosophical, and moral doctrines but by a pluralism of incompatible yet reasonable comprehensive doctrines. No one of these doctrines is affirmed by citizens generally."[15]

Such conflicts, as Rawls and Bohman agree, concern not merely interests but principles of a profoundly moral nature. They invoke, as we have seen, divergent understandings of what normative terms such as consent, authority, rights, and responsibility mean in political practice. They also succeed in bringing people together who share understandings of these concepts as they are highlighted in risk politics. Thus, they stand at the foundation of the curiously communal yet adversarial tone of contemporary politics. This is the second general conclusion to draw from our examples: risk feeds the identity politics of cultural pluralism, threatening the disintegration of liberal society within the pursuit of communal unity. John Gray calls the new politics of liberalism "agonistic liberalism." It is still liberalism because it is a politics based on contestations between equal members of society. It is agonistic because the conflict is not only over interests but also "ultimate values," even identity itself.[16] We will return to this conception later in this chapter.

The complexities of identity politics involving claims of minority status, ethnic neighborhoods, gender differences, and native languages that are convulsing both Eastern Europe and US cities cannot be explored in their entirety here. But, as they also come into play in risk issues, we should note why they threaten liberal politics so. Because liberalism begins with a principle of tolerance for cultural difference, it presumes that such differences can somehow be politically mediated. But as Rawls, Kymlicka, and others have made clear, identity politics is not the fruition of liberalism but potentially its demise. Why is this? The answer can be seen within risk politics. The plurality of belief and principle extant in risk politics and its disputes demonstrates how risk both unites people into a shared identity and set of principles partly defined by a shared relationship to some risk, while at the same time threatening the disintegration of the larger community (e.g., the nation) into balkanized groups and localities warring over siting, definitions of life, research priorities, and a host of other risk-related but only partially technical decisions. Risks build collective identities that demand public recognition and accounting, demands that cannot be fully met by the liberal acknowledgment of a private right of expression or toleration for its products. More is at stake here than interest, and the rights that protect them. Risk politics is a struggle for identity.

Finally, the third conclusion to draw from our examples of identity issues operating within risk politics is that we are indeed talking about politics here. The exercise and pursuit of power take place on several levels within risk politics, all of which go into the creation of self-identities of relevant interested groups. From appeals to provide money for research (or to withdraw funds for certain kinds of research) to seeking legislation, regulations, court decisions, or even presidential orders or vetoes, the institutions of government are all obviously in full operation in the politics of risk. And, of course, other mechanisms of political influence and power—from interest groups to civic associations to citizen blockades at the gates of nuclear power plants—are often in evidence as well in the swirl of risk decision making. To some extent risk politics is the very picture of politics-as-usual in a liberal democracy experiencing the turmoil that comes from its tolerant cultural pluralism.

On a more subtle level, however, risk politics is special and different in the threat it poses to liberal polities. As technological risk immerses citizens and the nation as a whole in a struggle for the identity of a community and its members, it exercises the power of creation—the making of selves and others through the definition of difference. Self-definition, whether of individuals, groups, or nations, invokes the power to include and exclude, to define what is the same and acceptable as opposed to the other and unacceptable. Such is the nature of identity politics, as Connolly explains:

> Power plays a prominent role in this endless play of definition, counter-definition, and counters to counter-definitions . . . [for] power is always inscribed in the relation an exclusive identity bears to the differences it constitutes. . . . The paradox has several dimensions or formulations. One is that if there is no true identity, the attempts to establish one as if it were true involves power, while if there is a true identity susceptible to realization, the attempt to pluralize and politicize identities militates against achievement of the highest good.[17]

We have seen how risk participates in and exacerbates the power-ladenness of identity politics and does so for individuals, groups, even whole societies struggling to assert their self-definition. Because identity has become so saturated with the exercise of power, liberal politics can no longer pretend that the exercise itself can somehow be contained in the realm of private individuals pursuing their interests and taking whatever risks are dictated thereby. We now realize that the whole idea of a purely private risk has been invalidated by the emergence of modern technological risks. The same is true for identity—risks have made identities public in evincing the play for power that risk politics invariably displays. Thus, identity in the society of emergent risk is itself emergent; it is the identity not of private persons but of citizens.

Risk and Emergent Citizenship

The meaning of citizenship is a wedge separating classical from modern political thought. For the classics, especially Plato, citizenship was a source—for virtually all individuals really the only source—of identification with an idea (and reality) of the good that was larger than oneself. Citizenship provided individuals with meaningful lives of public participation within a politically defined sphere of virtue and order. For moderns beginning with Hobbes, citizenship has less to do with self-identity, virtue, or even happiness and more to do with a political guarantee of the opportunity to seek happiness (or possibly virtue), most often in the private areas of life rather than the political. Thus, modern citizenship concerns the absence of interference with the individual ability to seek one's own vision of the good life.

The modern view of citizenship beginning with Hobbes and continuing with some interruptions (notably Rousseau) into contemporary Western societies is sometimes viewed as an impoverished one in comparison with the classics. Centered as it is on negative liberty—on the inviolability and autonomy of the private realm—modern citizenship incorporates a partial denial of politics and of the worth of political activity. Michael Walzer summarizes the liberal view of citizenship in terms of the shared political liberty that promises protection:

> But this search for protection assumes the primacy of what is protected, namely the private or familial world. It is there that men and women find the greater part of their happiness; they "enjoy" protection but rejoice in something else. They are not political people; they have other interests, in salvation or business or love or art and literature. For them, the political community is only a necessary framework, a set of external arrangements, not a common life.[18]

Since modern citizenship does not define what should be chosen as much as it protects the liberty to choose, Walzer concludes that it is "unlikely to be the primary identity or the consuming passion of men and women living in complex and highly differentiated societies, where politics competes for time and attention with class, ethnicity, religion, and family." Added to this list of the political realm's competitors, the biological, medical, and political groupings identified, as we saw, by modern risk also, as Walzer claims, "do not draw people together but rather separate and divide them. Separation and division make for the primacy of private realm."[19] In its encouragement of identity politics, then, modern risk seems to feed the modernist denial of political community and of the meaningfulness and virtue of citizenship as a positive identification with a larger, public, common notion of goodness.

Still, in its creativity in the creation of groups of people identified by subjection to a shared risk, modern risk does not lead individuals back to the strictly private realm. Instead, the identity politics resulting from risk brings people together in their antipathy to authoritative decisions and policies emanating from public institutions. In a sense, NIMBY politics is communal in that it unites people in a civic endeavor to fight risk. Yet it is also exclusionary in its insistence that the public as a whole, with its institutions and interests, constitutes the opposition. Furthermore, as we have seen, much of the politics of modern risk centers on arguments claiming rights as trumps in order to protect the inviolability of our private worlds as symbolized by our metaphorical—and sometimes literal—backyards. Thus, in its evocation of identity politics, risk both defends and defeats the liberal understanding of citizenship.

Within liberalism, citizenship is defined largely by the guarantees of independence and private liberty that rights-as-trumps imply, because these are of paramount importance in protecting the insularity of the private realm. Thus, Conover, Leonard, and Searing construe liberal citizenship as embodying two elements: first, the guaranteed liberty providing that we "may claim for ourselves particular privileges—in the idiom of liberalism, 'rights'—against interference in our choices from other individuals, society, or government." Exercising this liberty does not empty the concept of any notion of responsibility, but, they admit, responsibility as the second element of liberal citizenship contains only a duty "to maintain the rights of others," requiring "no action beyond those minimally required for the preservation of individual autonomy."[20] Such a view of responsibility supports Walzer's comment that within the liberal community, there are no citizens "out there" other than "rights-bearing, voluntarily associating, freely speaking liberal selves."[21]

A provocative application of this liberal view of citizenship to the whole area of technological risk is Philip J. Frankenfeld's concept of "technological citizenship." Frankenfeld argues that risks today have defined a "new social contract of complexity," within which citizenship presumes membership in a more global "technological polity," whose boundaries are defined not by national borders but by "the impacts of a technology or technology in general."[22] Frankenfeld concludes that the broadened realm of the political sphere requires that the rights of citizenship must be expanded beyond the traditional liberal conception to include

> equalizing the possession by all people in all generations within this polity of the political resources they need to protect themselves from and verify their safety amid complex, concealable environmental hazards, such as toxic chemicals and wastes, nuclear power and wastes, and recombinant DNA. This equality in self-protection and self-verification of safety applies to all points in an environmental hazard's advent, siting, use, detection, and remedy.[23]

Frankenfeld's concept forcibly expands the scope of liberal citizenship to include a new area of political reality—modern risk—and even makes the concept globally applicable, compensating for Ignatieff's concerns. Clearly, he too acknowledges the creative power of technological risk in its impact on traditional liberal concepts. Yet, beyond this widening of citizenship's circle of putative relevance, Frankenfeld's "technological citizenship" offers only a small emendation on the liberal idea. Missing still is a recognition that risk alters the understanding of citizenship by invoking individual duties and obligations that cannot be fully satisfied within the private realm or by a mere recognition of the rights of others. Obligations are implied by technological citizenship, Frankenfeld posits, but he admits that they are "minimal" and require merely that one use knowledge for "one's own self-validation of safety and peace of mind" and for political participation in determining the majority will on issues.[24]

Liberalism's view of citizenship is not up to the rigors modern risks pose, even, as in Frankenfeld's case, when liberalism is forced to acknowledge the conceptual power of risk. Furthermore, as contemporary versions of liberal citizenship, views such as Frankenfeld's and Conover et al.'s in their belittling of the duty-content of citizenship, are themselves something of a denial even of the classical liberal view. Certainly Locke understood that the ability of individuals to recognize without tutelage their moral responsibilities was a major reason legitimating their claim to liberty. The capacity to acknowledge and act upon moral obligation is a natural attribute of individuals in evidence for Locke in the state of nature and gives force to the laws of nature, thereby ensuring that the natural state is relatively peaceful rather than embroiled in the Hobbesian anarchic nightmare. That individuals tacitly consent to the laws of nature in the state of nature is also Locke's ultimate basis for legitimating tacit consent as the basis for political obligation in the liberal state, as we saw in chapter two. And although Locke's approach means that political obligation dissolves into a species of moral duty—ultimately to the detriment to the full development of political community—it is clear nonetheless that Locke appreciates, even more fully than his inheritors today, the place that duty holds within the logic of liberal moral theory and the politics constructed upon it. Originally, as an argument for individual liberty, liberalism grounded politics on the capacity for moral autonomy willingly exercised by all individuals, regardless of the exercise of power forcing them to adhere to their obligations (as in Hobbes). This autonomous ability to exercise liberty *and* to recognize duty is, for Locke, the grounds for citizenship.

But even if contemporary theorists like Frankenfeld and Dworkin draw the liberal acceptance of duty too narrowly to claim even a strictly Lockean paternity, liberalism from Locke onward nevertheless presents three aspects of moral duties that make the version of citizenship constructed from them (in combination with negative liberty) insufficient to meet the demands of contemporary risk. First, with few exceptions, liberalism speaks only of private duties—obligations owed others as a matter of private contract or debt. Also, duties are to other individuals, not to society as a whole or to an idea (or ideal). Because society is viewed by liberalism not organically but as an aggregate of individual people, all duties must have recognizable and individual recipients to whom something is owed as a matter of right. Thus, if I have wronged someone I am responsible for the harm and have a duty to make it right, but I cannot be forced to give something up in the name of some larger good for which I have not contracted or whose recipients I have not myself harmed. Therein lies liberalism's difficulty with the NIMBY syndrome and with claims of justice for future generations.

Public duties, such as an obligation to participate in politics—even to vote—are not part of liberalism's moral code. They cannot be, for that would require a commonly held and substantive version of public good that goes beyond the private duty to obey the law and keep the peace. Participation as a duty can be spoken of only by liberals as it adheres in the private realm, within relationships that are the results of freely negotiated consensual contracts (e.g., marriage or business arrangements). This rejection of public duties is the point of divergence between liberals such as Locke and civic republicans such as Rousseau and contemporaries Sandel and MacIntyre, all of whom view liberty less negatively and allow for various forms of "forcing to be free." The civic republican view is considerably more of a communitarian understanding of what it means to be a citizen. Citizenship under such a "civic" interpretation requires the recognition of public identities, both of self and of others, either individually or in groups. It is a view of citizenship, as Elshtain says, as a "*civic* identity, not primarily a private sinecure"[25] (emphasis Elshtain's).

Second, liberalism defines relationships of duty and moral obligation as relying strictly on causal features. As we saw in the last chapter, this means that responsibility for harm can entail only if there is a clear causal connection between actor and consequence and if the consequence is concretely felt by identifiable persons. On a broader level, consequences of collective action (or inaction) ranging from racial discrimination and prejudice to policy decisions resulting in technological risks (e.g., development of nuclear power) impose no moral or civic duties whatsoever on any single individual, unless that person is specifically identifiable as having been a cause of the harm. Obviously, many social problems, in addition to risk issues, today require different, less reductionist approaches to the idea of duty if solutions are to be found.[26]

Third, liberalism conceives of duties as the concomitants of rights and of the right, not of the good. In other words, duties are the other side of the rights coin—if someone has a right to something, then someone else has a duty to perform or refrain from performing some correlative action. Whether one believes the rights to be "natural" or not, they imply that someone else has duties (also natural, not in the sense of being owed to nature but to some other person endowed with natural rights). Similarly, in civil society duties along with rights are encapsulated in a correct structuring of the social and political realm—in a just (or for Rawls, a fair) system of institutions and laws. That is, duties are the product of the "right," in terms of correct or fair procedures that protect rights. Duties do not descend upon individuals from a conception of goodness that obligates, such as universal moral standards, images of the divine, or even collectively approved ultimate goals. Such notions of the "good" liberals suspect of challenging the sanctity of individual rights.

As Mary G. Dietz points out, it is this disagreement concerning the efficacy of the good, and of what duties are derived from it, that separates liberals from communitarians. All communitarians agree, she claims, that "the principle of *the common good* outweighs the principle of individual rights and duties."[27] The liberal/communitarian debate of recent years is really a debate about autonomy as a value of liberal theory. Communitarians such as Sandel, we have already observed, criticize liberalism for ignoring the connectedness of individuals—and their need for it—in its drive to validate the autonomy of every person. It is the liberal view of identity that is at issue here: can individuals define themselves (and their "good") autonomously in their private worlds and attachments, and should they, as a matter of right? Or are all persons "encumbered" to such an extent with shared understandings, and born into roles defined largely by external relationships with names like "citizen," that the whole idea of autonomy is both pernicious and chimerical anyway? Or, as a third possibility, is there a necessary connection between public and private identity that cannot be broken? Connolly assumes this last to be the question of the moment in defining what democracy is to become:

> [O]ne's self-identification as a citizen in a democracy, where one's choices and judgments matter in the public realm, informs the orientation one takes to a variety of other social roles such as those of spouse, lover, student, worker, taxpayer, and consumer. One is likely to be more independent, more assertive about one's rights, more disposed to respond politically to grievances or injustices in these domains if one has interiorized the definition of oneself as a citizen in a democratic state.[28]

Making the public self-definition private, as Connolly suggests, is recognizable as the Rousseauian solution to the "social problem" of being born free yet living encumbered by society's chains. Democracy and democratic citizenship are the elements of this solution, but both imply a much broader set of duties to others and to a shared sense of value than liberal definitions. Autonomy in a strictly liberal sense is thereby jeopardized but at the same time is reborn in new, social garb once democracy is embraced. It is Rousseau's paradox of freedom and duty encapsulated in the vision of the free citizen "flying to the public assembly."

Connolly (with Rousseau) is correct in recognizing that the tensions underlying identity and difference in modern society can be negotiated only democratically. This is especially true for the problems raised by the identity-affecting power of modern risks. As we have seen, risks have a way of forcing identities upon us, of thrusting us into groups characterized by shared vulnerabilities not of our own choosing. In our shared vulnerability we recognize the incompleteness of our liberty individually to define our own private realms and our identities within them. That is, we recognize that liberal autonomy is not possible. Liberal solutions such as relying on rights-as-trumps at this point are demonstrably wrongheaded and lead only to the conflict of interests and subsequent policy stalemate characteristic of many risk decisions. What is needed is an understanding of citizenship that recasts autonomy and liberty, and therefore identity as well, as the social constructs they really are.

Such a view of citizenship capable of providing solutions to the problems of modern risk without compromising democratic liberty must take into account the impact of risk on individual identities and individual realms of action. As we have witnessed throughout this book, the general consequence of modern risk is to dissolve distinctions between public and private, between individual and collective identities, between mine and thine. As modern risks are emergent phenomena, they call for collective responses to privately felt threats and for the recognition of public duties in the pursuit of private ends and rights. Thus, risk demands that the citizen be defined not in terms of the privileges guaranteed by the state for the pursuance of private goals, or even in terms of the overall relationship of the individual to the state. Rather, citizens are persons discovering their identities through their relations with others, relations (including those dictated by risks) characterized by actions in response to risk that define a public discourse. These relations presume new responsibilities for individuals as civic persons. Citizenship requires the recognition of individual responsibility for an emergent vulnerability that one alone has not caused. Thus, citizenship may claim a duty of participation from individuals that will take time, effort, even sacrifice. Citizens must become active participants in a public discourse about risks and responsibilities, since it is within this discourse that solutions can be found for risks, in the pursuit and exercise of collective political judgment. As a preliminary summary, then, modern risks demand "emergent citizens," whose identities are defined not by the private characteristics of individuals (e.g., reason, rights, etc.) but by their civic relations with others; whose political actions are the manifestation of those relationships as they relate to risk, both inside and outside present political institutions (thereby potentially creating new institutions); and whose collective deliberations arrive at decisions that elevate collective judgment rather than technical expertise as the authoritative voice within risk politics.

In its rejection of abstract qualities such as reason and solely private attributes such as rights as the basis of citizenship, this emergent view clearly departs from liberalism but not, I will argue, from the idea of democracy. Emergent citizenship presumes, first, that citizenship defines a type of relational activity rather than a type of private guarantee and, second, that citizens emerge from private individuals through the exercise of autonomy within a civic context. Following French psychologist Jean Piaget, Mary Gibson provides a useful framework for such a view:

> Autonomy, on this conception, is not a fundamentally individualistic idea that must somehow be accommodated to the undeniable interdependency of individuals with one another and their environment. Rather, autonomy is, in its origins and at its very core, a social phenomenon resulting from relations of cooperation and attitudes of mutual respect among persons who regard themselves and each other as equals.[29]

Autonomy for Gibson is not, as liberals such as Locke and Mill assumed, a set of capacities characteristic of individuals viewed in isolation, displayed either in a state of nature or a solitary utilitarian calculus. Rather, as de Tocqueville urged, it is those capacities for reflection and critical judgment *as they are exercised* within social contexts—as Gibson states, "the degree to which one exercises autonomy is the degree to which one is autonomous."[30] In the context of risk decisions, then, autonomy is not threatened by the emergent nature of the risks themselves, even though that nature demonstrates both how risks threaten individuals and link them to others in unexpected ways. Rather, risks threaten the autonomy of individuals only if individuals do not respond with autonomous action. That action, in turn, must be the action of citizens: individuals whose autonomy emerges through their political relations with others, as they make collective decisions about risk. The institutional framework of those decisions—and their nature as democratic—I explore in the next section; here, we need to delineate the identity of citizens as autonomous, yet socially defined individuals.

In rethinking democratic theory, Carol C. Gould lays a basis for seeing individuals confronting risks as potentially autonomous citizens. Her foundation begins by viewing the nature of democracy as developmental, as grounded in the positive liberty for citizens to develop their potential as individual human beings. Democracy for Gould, then, is the institutionalization of "the principle of *prima facie* equal rights to the conditions of self-development."[31]

But the holders of those rights are individuals seen somewhat differently by Gould than by traditional liberal thought. As social beings, individuals require relations of reciprocity as one of the social conditions of their self-development. This reciprocity is necessary for both the physical and psychological resources that individuals need for achievement of their purposes. So significant is the need for reciprocity, Gould argues, that individuals should be seen ontologically in a way that is subtly—but significantly—different than in liberal theory. Persons exist, Gould concludes, as "individuals-in-relations," whose

> characteristic mode of being, that is, their activity, is relational or essentially involves their relations with others. Thus these individuals act fundamentally in and through social relations. The individuals are therefore ontologically primary, but the relations among them are also essential aspects of their being. However, these relations do not exist independently or apart from the individuals who are related. Rather, they are relational properties of these individuals.[32]

In Gould's ontology of individuals-in-relations, she is careful not to say that all individuals are reducible to the sum of their relationships, even though, for instance, the identity of a teacher can exist only in relation to students. Individuals bring an original purposefulness and choice to the relations they initiate with others that may not be part of the relationship itself. Thus, Gould avoids a completely social definition of individuality. Yet the social element of her definition is clearly more pronounced than in classical liberal views, either of the state of nature theorists or later utilitarians such as Mill.

As an argument for democracy and for the positive liberty it offers individuals, Gould's influential rethinking of the democratic individual offers important insights for us as we consider the impacts of risk on individual identities and the lives of citizens. As we have seen, almost nothing calls attention to the interconnectedness of individual lives as much as do modern technological risks.[33] And almost nothing seems to challenge our autonomy more, in the face of modern risks we feel anything but free as individuals to autonomously determine the future course of our lives. Risks entangle us in the lives of other people, in their presence, their interests, their bodies, their gene pools (and they in ours), to an extent that seems to render autonomy a chimerical concept of a long past, more innocently optimistic time. Yet modern risks are themselves the product of human action; risks are not themselves autonomous, merely collective: emergent manifestations of collective choice and action. They can be controlled, and controlled democratically, that is, in a way that protects the equal right of every person "who engages in a common activity . . . to participate in making decisions concerning such activity."[34] It is within these activities made common by the environment of technological risk that generates them, where control, even liberty, can yet be established by individuals jointly engaged in an attempt to take back authority over their risky lives. As individuals we are helpless in the face of modern risk; as reciprocally engaged public persons sharing an identity defined by modern risk, we can make decisions, take actions that are autonomous in a democratic sense of equality and mutual empowerment. As citizens in a democracy, that is, we can reclaim liberty and power over our collectively risky lives.

The Democracy of Risk

Several contemporary theorists of democracy have something to offer emergent citizens looking for ways to reassert authority over the technological risks that characterize modern political life, and it is not my intention here to summarize their contributions. Elements of "strong" democracy (Barber), "discursive" democracy (Dryzek), "ennobled" democracy (Pangle), "deliberative" democracy (Fishkin and Guttman), "civic" democracy (Elshtain and Sandel), and "agonistic" democracy (Connolly) illuminate the two fundamental questions that democratic politics must answer in response to the challenges posed by modern technological risk.[35] First, what does the politics of risk require to arrive at democratic solutions to risk issues? Second, what does supplying those requirements demand of citizens and political institutions?

To begin to answer these questions, we must restate the lesson gleaned from our exploration in previous chapters of the variety of risk policy disputes encapsulated within the politics of risk. That lesson is simply that a political process perceived as a combination of the pursuit of individual interests with the application of technical expertise and advice will not lead to a resolution of most risk controversies. In fact, this convergence of conflicting interests and expertise has most often aggravated the policy process rather than ameliorated it.

The reasons behind this policy failure are both ontological and epistemological. On an ontological level, we have seen that the nature of risk as an emergent phenomenon is not amenable to a politics based on a set of reductionist conceptions of interest, consent, authority, rights, and responsibility. Instead, we should expect that solutions to emergent problems are likely to be themselves emergent. Second, on an epistemological level, the reductionism of liberal politics leaves policy makers simply at a loss about how to proceed in confronting risks. The knowledge system upon which liberalism is based proceeds along an instrumental understanding of rationality presuming "that complex phenomena are best understood through intelligent disaggregation into their component parts."[36] Because emergent phenomena risks cannot fruitfully be so deconstructed, liberal policy makers (as well as citizens) are epistemologically at sea in their approach to risk.

Two alternatives present themselves as possible candidates for reconciling the ontological and epistemological incongruencies of risk and liberal politics. As Dryzek, Barber, and Pangle understand, they both incorporate a view of rationality upon which an effective politics can be based. Only one of these views of rationality capable of coping with risk is amenable to democratic institutions, however, as well as to the overriding goal of individual liberty and social autonomy. The other alternative—not so democratically disposed—is what the ancient Greeks saw as the unique ability of the legitimate statesman to see beyond individual interest and the reduced singularities of politics to the larger picture of ultimate ends and holistic phenomena, realities, and virtues. Such individuals exhibited the practical judgment and critical reason (*phronesis*) that identified them as Plato's philosopher-kings, Aristotle's statesman, and even Rousseau's Great Legislator.[37] For the Greeks, only these epic figures should make decisions with the special ontological and epistemological demands of risk politics, for only they could view politics and its objects in their emergent realities, unsullied by the contaminating narrowness of error, partiality, or self-interest.

Leaders with these qualities have never been exactly thick on the ground throughout political history, as all their apologists have understood. Besides, it is not at all clear that the capacities they exercise and the rationality they express can fit into a decision process in any way describable as democratic. Though Aristotle and, today, Hannah Arendt and Thomas Pangle invite all citizens into "participation in the theoretical life," in our time it is science and its practitioners who have most often claimed the legacy of the philosopher-king and also therefore the wisdom to make political decisions.[38] This claim is the basis of the argument that science must have complete autonomy over its own governance and in decisions concerning how its products are put to social use. Such claims of autonomy have more often caused policy disputes over risk than resolved them.[39]

If the Greek ideal of practical wisdom is not readily forthcoming to provide the epistemological readiness to deal with risk, is there another alternative? The answer is yes, but it is an alternative less rooted in presumptions either that there is a truth to know, or a way (or person) to know it, than in the view that politics is not about truths as much as about collective judgments concerning the public good. Benjamin Barber offers his conception of "political judgment" as the democratic alternative to the elusive *phronesis*. Grounded in a pragmatic denial that theory is separate from the realm of practice, political judgment is "public thinking," where the emphasis is upon the political, "public" aspect rather than the cognitive, "thinking" element. Barber says that political judgment merges public judgment and collective action in a democratic discourse:

> My argument draws on the Anglo-American experience with praxis and looks to theorists of political action such as Burke, Tocqueville, and Dewey for justification. It construes political judgment as a function of commonality that can be exercised only by citizens interacting with one another in the context of mutual deliberation and decision. Solitary individuals may make moral or aesthetic judgments, solitary judges sitting on the bench may render legal judgments, but citizens can produce political judgments only as a body acting corporately. . . . Political judgment is something produced by politics rather than by cognition . . . [and] political judgment is the multitude deliberating, the multitude in action.[40]

As Barber notes, political judgment is uniquely suited to democracy, because it "maximizes interaction and thus guarantees the diversity and generality that is crucial to prudent judgment."[41] It also clearly presumes a broadening of political participation and therefore requires alterations in the political institutions of most contemporary democratic systems. Before exploring those, however, we should also realize that, specifically in political deliberations involving risk, political judgment is not at all the same as technical expertise. In fact, although judgment within Barber's design of "strong democracy" does not necessarily replace expertise in political decision making, it supersedes expertise just as the voice of citizens supersedes that of technical expertise within a strongly democratic risk policy. "Expertise," concludes Barber, "may facilitate judgment, but it does not constitute judgment."[42] Risk decisions need to be informed by expert advice, but in the end they raise the same issues about resource allocation and risk acceptability as do all political questions. From the perspective of democratic decision making, risk issues are no different from other political choices. Therefore, in Barber's view, risk decisions must emerge from the same strongly democratic processes that produce political judgment in other areas.

Political judgment as construed by Barber seems especially apt as a goal of risk politics because it expresses the same emergent character as risks themselves do. A similar approach is taken by John Dryzek in his design for "discursive" democratic institutions. Following Habermas, Dryzek views liberal politics as the domination of instrumental rationality in the service of "disaggregated interests," leading to the loss of community, democracy, individual liberty, and spontaneity, and to the creation of a bureaucratic system that stultifies real political discourse and, significantly, "makes effective and appropriate policy analysis impossible."[43]

In opposition to the repressiveness of instrumental reason and the politics of domination it generates, Dryzek offers what he calls discursive rationality, roughly Habermas's idea of "communicative rationality" applied to the possibilities unleashed by contemporary communications technologies. Dryzek does not deny that those technologies are themselves the developmental products of the application of instrumental rationality in the pursuit of profit and domination and, as such, and as we have seen, pose risks of repression and uniformity. Yet

> these same developments also enable discursive and critical reconstruction of the lifeworld. Subjects can coordinate their actions through talk rather than unthinking compliance with norms, and it is through talk that an attack can be launched against instrumental rationalization. To date, instrumental rationality has got the better of this conflict . . . yet resistance based on reflection, talk, and communicative action remains a possibility.[44]

Such resistance is the basis of discursive democracy, the application of communicative rationality to "social problems (especially complex ones), political and policy-making institutions, policy analysis, and empirical political science." Within this new politics of talk, "the only remaining authority is that of good argument."[45]

Both Barber's and Dryzek's views of democratic policy making share a belief in the necessity of a much enlarged realm of citizen deliberation in decision making, especially in areas of risk politics that encompass especially complex interconnections between the lives of individual citizens. Both see this necessity as a response to the ever growing threat to individual autonomy posed by modern technologies (as the products of instrumental rationality) and the risks they present. Neither Barber nor Dryzek fully realizes, however, that it is the ontological status of modern technological risks that makes the political transformations they seek so apt. Emergent risks can be confronted successfully only by means of a collectively rational judgment that itself "emerges" through a free process of talk and mutual exchange. Any other rationality or judgment risks either a fallacious reductionism or an unacceptably undemocratic claim of authority based on superhuman wisdom. That is to say, the emergent nature of risks can be confronted *democratically* only by adopting the discursive design for institutions that Dryzek proposes, institutions that operate as the amphitheaters within which Barber's political judgment emerges.

I deliberately choose the image of amphitheater as the locus of democratic political discourse about risk, because the identity-affecting nature of modern risks guarantees that more than just interests will be in conflict in the discursive battles over risk policy. Indeed, democratic participation in our technological age will little resemble Rousseau's calm deference to the general will that emerges in the local assembly. Rather, risk politics engages our self-images and identities to an extent that political discourse means the assertion not just of what one wants or needs (however interests are construed) but of what one values and perceives oneself to be. This is not politics merely as the conflict of interests, but the conflict of selves, an agonistic politics in which the struggle is not only to get one's way but also to recognize one's own identity in that of diverse others and vice versa and to do so sufficiently to enable a view of the political world—and one's identity in it—as it really is in its emergent reality. Democratic politics is an agonistic search for self that, as Connolly describes, "expresses and unsettles" the connections between "personal and collective identity":

> Democratic turbulence disturbs established commonalities: it shows them to be complex contrivances; it brings elements of contestability within them; it exposes possibilities suppressed and actualities enabled by contestable settlements. By fos-

> tering the experience of contingency and relationality in identity, democracy disturbs the closure of self-identity and, sometimes, provides a medium for modifying the terms of collective identity.[46]

As we have seen, nothing in politics reinforces the experience of "contingency and relationality in identity" as much as do modern risks and the policy disputes surrounding them. As such, nothing holds out more possibility for the reconstitution of democratic politics—in all its raucous possibility—than do the risks that threaten our identities, our lives, our futures. Political theorists like Connolly, Dryzek, Gutmann, and others have envisioned it; our recognition of the frightening efficacy of modern risk can now make it happen.

What then is required of citizens and their political institutions to make a truly democratic politics of risk a reality? A sizable and growing literature already exists in answer, but some points are worth emphasizing. First, as Dryzek notes, a truly communicative democratic discourse requires "communicatively competent citizens."[47] At the very least, competency will necessitate access to more resources, time, and information than many citizens possess at present. Competency will mean, then, additional participation by citizens in the deliberative process for risk policy, and levels of participation are determined as much, if not more, by these variables than by mere interest(s) alone.[48] In other words, truly democratic risk policy requires a democratic equality in more than just voting rights and economic opportunity. This is a vast prerequisite, of course, but the technologies themselves that produce risks and therefore demand policy also offer, as in the example of modern communications, new opportunities for equalization.

Communicative competence for risk issue participation does not require, as noted above, actual technical expertise. Yet it does presume engagement in the public realm beyond what is presently typical of liberal citizens who construe democracy's guarantee of liberty in excessively negative fashion. Decision making for risk demands, in short, the consent of citizens, both as a matter of their own autonomy and as a requirement for policy success. But, as we explored at length in chapter two, consent for risk cannot be tacit; only an expression of deliberative assent is acceptable both for citizens and for policy makers aware of the consequences of the NIMBY wars. Such expressed consent implies activity on the part of citizens;[49] thus, it is an imposition into the private realm. But what most arguments for a newly participatory democratic practice ignore is that participation is an imposition based explicitly on consent. And it is real, not tacit, consent that expresses itself explicitly through action and competent democratic communication. It will take time and effort to become communicatively competent, but living with risk ensures that the definition of liberty can no longer be purely negative—it cannot countenance withdrawal from the public realm.

Public involvement in risk issues is thus an inescapable duty of democratic citizens, a responsibility born of the vulnerability both of self and others that comes from postponing for too long decisions about risk. In some cases, the delay has already reached dangerous proportions, as the risks persist and multiply within the policy void surrounding nuclear disposal technologies as well as several medical research and reproductive technologies. Furthermore, this duty to participate may at times demand the sacrifice of more than just time on the part of some citizens, as in the case of necessary siting of hazardous installations. In these cases the responsibility is neither easy nor equally distributed. Citizens so affected have the right to expect compensation from a grateful community, but as we saw in chapter five, they do not have the automatic right to deny the community's will. The rights of citizens do not extend to the trumping of proposed risk policy; they do guarantee that every voice will be heard and its speaker respected. Compensation is both the acknowledgment and the measure of that respect.

The duties of citizenship invoked by risk are undeniably an imposition, and because they are it is clear that something more than the liberal understanding of citizenship is called for to live freely in our risky present. Amy Gutmann clarifies the point:

> Liberalism prohibits collective choices from interfering with personal freedom. Democracy promises that they reflect popular will. Two competing ends cannot simultaneously be maximized. Democracy and liberalism part company when collective choices threaten to interfere with personal freedom, or personal freedom threatens to interfere with collective choice.[50]

The democracy of risk presents a serious challenge to liberalism's definition of citizenship, since at the very least it requires that citizens recognize the diminution of their private realm and its replacement with a far greater realm of public engagement. At times the pervasivness of risk might seem even to dissolve the line between public and private life completely. In reality, risk makes drawing the line between public and private itself a matter of contention and political deliberation, thereby placing liberalism's most basic conceptualization at risk.[51] Yet this challenge does not deny that liberty is of prime value within a democracy that deliberately—and deliberatively—confronts risk. It is only through the collective effort necessary to make decisions that a new kind of freedom can be achieved regarding risk, a liberty not from all risk but from the fear and uncertainty (not to mention dangerous consequences) that risks instill when they are not allowed into the realm of public discourse. Also, this is a liberty that strives for autonomy not through isolation or through abstract attributions of the private cognitive abilities and metaphysical rights of individuals. Instead, the liberty attained through the democratic negotiation of risks is a positive liberty that leads to social autonomy through action in the world. It is truly political liberty, "the freedom to judge, to choose, to act when the grounds of choice are not given a priori or by fiat or by impeachable cognition. To be political is thus to be free with a vengeance—to have to make judgments without guiding standards or determining norms, yet under an ineluctable pressure to act."[52]

But are citizens reared in liberal ways capable of shouldering the responsibilities implied by such liberty, such autonomy, such action? Barber's and Dryzek's views of citizenship may seem overly heroic in estimating the level of uncertainty under which they expect citizens comfortably to act. But we should make some disclaimers here about what citizens must know and do in order to participate autonomously in risk decisions. First, as stated before, expertise is not expected of them. Second, though we have seen how most risk issues powerfully invoke complex questions of moral and political theory, it is unfair to expect that citizens must then become moral and political theorists in order to prove themselves collectively competent to make risk policy. Such expectations often underlie the mistaken belief that citizens do not measure up to the demands of collective life. Donald Moon points out this error endemic to much democratic theorizing:

> The traditional formulation of the problem of citizen competence tacitly incorporates the view that the activity of citizenship and that of political theory or philosophy are in crucial respects the same. As private persons, we seek to make correct choices regarding the moral issues we face; as citizens, we seek to participate in deliberations on public issues to reach rational, well-reasoned judgments.[53]

If, as Barber and Dryzek insist, it is not necessary for citizens to become experts in any sense before their contribution to political discourse has value, then clearly this includes achieving the qualifications of moral or political philosopher. Citizens cannot be expected to approach a particular issue or dispute as if its particular facts or risks must be generalized to fit all possible cases before they are allowed to pass judgment on it. Most moral reasoning and political disputes engage citizens as they perceive the dispute's immediate impacts and risks, and it is to these that citizens respond. Does this narrowness of perspective make citizens incompetent deciders of important issues? Moon responds that, in Isaiah Berlin's famous cognitive distinction, it only makes them "foxes" rather than "hedgehogs":

> The problem of citizen competence, I would like to suggest, is not so much that citizens fail to be theorists as it is that much of this discussion is premised on a mistaken conception of theory in which only a "hedgehog" [one who successfully relates everything to a single, universal organizing principle] style of theorizing is allowed.[54]

Risk policy requires only that citizens be "foxes," pursuing many ends not always completely logical in their relation to one another and investing in no single moral or aesthetic principle.[55] Since their identities are deeply engaged in risk disputes, and multiple risks can create multiple identities, ideological consistency and even logic cannot always be expected in citizens' responses. But, as Moon concludes, this means only that political discourse must be tolerant, open, and aimed at political learning by citizens, not that it waits for its success upon citizens successfully emulating the cold logic of moral philosophy. Citizens need not become experts of either technology, morality, or philosophy, but learning is required, and this is a demand both on institutions and citizens.

A third disclaimer about the capacities of citizens to make risk policy contributions is that the autonomy they seek through political action is not inconsistent with either the existence of political leaders or even delegation of authority to them. Deliberative democracy guarantees citizens the opportunity to collectively consider issues, yet, as Gutmann points out, unlike other participatory models, "deliberative democracy takes account of the burden of political action and the advantages of division of labor."[56] Deliberative democracy makes autonomy its aim, but, as Gibson notes, this is an autonomy found in relations and in discussions with others. Such autonomy does not presume popular rule in the sense of direct democracy; thus, institutions are conceivable that deliver the accountability that autonomy demands without requiring constant oversight by citizens.

For risk politics, many such institutions already exist or have been proposed. From public hearings to referendums, to various forms of mediation and regulation negotiation, to Dryzek's "discursive design" for institutions, there is no shortage of proposals for decision processes and policy organs for risk policy that invest authority in citizens' hands.[57] As democratically valuable as it is to try any and all of them, we should remember that the procedures of democracy are not the same as its substance. Rule *by* the people does not guarantee rule *for* them.[58] Of course, nothing but citizens themselves and their actions can provide that guarantee. And, of course, there is always the risk that even when all rule is by the people, they might still be wrong in their decisions and the results will not be in their interests. That is the nature of democracy—it is itself a risky business. But because it is, only in a democracy that emerges from the consensual actions of free citizens can we understand and hope successfully to confront the risks that emanate from their other businesses. In the final analysis, only democratic politics shares the emergent ontology of modern technological risks, but this is true only when citizens are actively engaged in the search for solutions to those risks. That is, only in emergent democracy can we expect to discover solutions to the problems of emergent risk.

Conclusion: The Political Community of Risk

Within the politics of risk resides the future of democracy; the same is true for the concepts that democracy embraces. The substantive meanings of consent, community, authority, rights, individual responsibility, liberty, and citizenship will be defined by their relevance within the procedures employed to make risk decisions. Given what we have seen of the impact of emergent risks on the traditional definitions of these terms of liberal political discourse, we would expect their new meanings to be less reductionist and individualistic than what liberalism stipulates. Yet, as political terms, they will retain their status as categories of meaning for liberal politics. This is to be expected since both liberal and risk politics rely on concepts of autonomy to reflect the significance of all other political concepts. Their divergence is in how that autonomy is construed—privately for liberalism, publicly and socially for deliberative risk politics. As Amy Gutmann hopefully concludes, a sufficiently deliberative model of democratic politics for risk can mediate this difference between liberalism and the politics of risk, because "deliberative democracy goes a long way toward reconciling democracy with liberalism by allying itself with autonomy, understood as self-determination through deliberation."[59] A different rendering of autonomy surely than that of traditional liberalism, yet it is one that, in its family resemblance to its liberal forebear, might through democratic deliberation smooth the transition to a new political reality.

Deliberative democratic mechanisms for risk decisions, because they retain an emphasis on autonomy, do not promise a tranquil political process. Quite the contrary, because risk decisions frequently demand choices between different and competing values as well as interests, they can be, in Gutmann's view, "agonizing." Indeed, the agonistic aspects of risk politics extend to the perceived identities of the political actors themselves, and we should expect conflicts of identity to be no less difficult in the choices they portend. Does this mean, then, that the final (or is it the first) political casualty of technological risk is the idea of community?

As we have seen, the communitarian effects of modern risks are somewhat paradoxical. Because risks affect not just interests but values and identities, they bring people together in a way that is potentially community-embracing. Indeed, technological risks sometimes even seem to create new communities, whether of the local victims of siting hazards, of hackers on the Internet, or of newly identifiable carriers of disease. In general, the impact of emergent modern risks is to reinforce the interconnectedness of individual lives; this seems a communitarian stimulus.

On the other hand, modern risks call attention to the different values, understandings, and identities held by whichever side of any risk issue one occupies. So risk politics puts (in some cases newly realized) groups of people ineluctably at odds with other groups to a degree that goes beyond merely the clash of interests. Thus, while wrenching individuals out of their private worlds into recognition of the risks they share with *some* others, modern risks, in the inequality of their distribution, seem to place numerical limits on *how many* others with whom one shares interests, values, identity.

In brief, emergent technological risks both unite and alienate. So what does this mean for the goal of community, so inadequately provided for by liberal politics, and, we might hope, better suited for the postliberal politics of risk? The answer lies, of course, in the future, as democracies and their citizens belatedly learn to cope with risk. But we might at least expect that although risk politics will never drive democratic citizens to a full communion of interests, values, and identities, it must—and does—of necessity drive them to a shared realm of action.

The galvanic impact of technological risks can be observed in virtually every daily newspaper, in stories covering the political activity stimulated by risks. Some of that activity is negative in its attempt to deny a social and collective response to risk. NIMBY politics is of this variety. Some of the political activity of risk is positive in its assertion of a recognized collective identity brought about by the recognition of risk and suggestions for response. This is the creative, civic dimension of modern risks, bringing people together in a shared perception of their likenesses as well as their vulnerabilities. The civic impacts of risk often lead to the political organizing activity of risk politics; sometimes it grows out of NIMBY claims, sometimes it precedes them, but it means that individuals have recognized their group interests, shared values, or collective identities sufficiently to begin taking action in response to the presence of risk. In every case, this activity focuses on the granting or withholding of consent to possible policy responses to risk.

As we saw in chapter two, liberalism relies excessively on tacit consent as a mode of legitimacy and political obligation and as a result stimulates a kind of moral community that allows rights to be exercised as trumps in a manner that arrests political decision making about risks. The moral community of liberalism allows individuals to remain in their private realms, disengaged from political activity as a matter of negative liberty, but retaining the right to prevent the community as a whole from acting. As risks undermine these hallmarks of political liberalism, they demand that citizens act, that they consider, deliberate, and express. In so doing, risks offer the opportunity to redefine the realm of consent within deliberative mechanisms of democratic politics. Risks make expressed consent a possibility.

Deliberative democracy requires mechanisms of expressed consent and, as the present politics of risk instructs, so does the very prospect of successful policymaking for risk. Thus, even though risk politics cannot promise—and, indeed, seems actively to prevent—the development of a community based on a set of shared understandings of life, value, and meaning (a moral community), it positively requires a community united in its will to act. That in itself is a community eagerly to be anticipated and sought. It is purely a political, not moral, community, defining its members and its understandings in pragmatic fashion as the need for decisions arises and in the very activity of making them. That activity presumes no commonality shared by its participants other than the action they in fact collectively participate in—the deliberation for decision. But that activity itself defines a community, the deliberatively democratic, *political* community of risk.

Notes

Introduction

1. William Leiss and Christina Chociolko (1994), *Risk and Responsibility* (Montreal: McGill-Queen's University Press), p. 6.

2. Though, as Jeffrey C. Isaacs argues in "The Strange Silence of Political Theory," we as political theorists have not written much about the remarkable events in eastern Europe and what was once the Soviet Union. I take Isaacs's critique seriously and agree that political theory has lately been on a self-possessed, even narcissistic journey of self-discovery. This book is an attempt to bring theory back into everyday politics and vice versa; along the way I expect the reader to (re)discover how difficult it is to locate the theory-ladenness of political events. See Isaacs (1995), *Political Theory* 23:4, pp. 636–52.

3. Alasdair MacIntyre (1966), *A Short History of Ethics* (New York: Macmillan), p. 3.

4. See Terence Ball, James Farr, and Russell L. Hanson (1989), *Political Innovation and Conceptual Change* (Cambridge: Cambridge University Press), p. 2.

5. Isaiah Berlin (1969), "Historical Inevitability," in *Four Essays on Liberty* (Oxford: Oxford University Press), p. 113.

6. John Dewey (1929), *The Quest for Certainty: A Study of the Relation and Knowledge and Action* (New York: Minton, Balch).

1. Risk and the Redefinition of Politics

1. We should keep in mind here and throughout the following chapters that though liberalism is still the dominant model of democracy, it is no longer recognized by political theorists as the only model. For a critique of excessive reliance on liberal concepts to define democracy, see the work of C. B. MacPherson, especially his (1996) *The Real World of Democracy* (Oxford: Oxford University Press).

2. I explore in later chapters and in greater detail the relevance of the individual political theories of Locke, Hobbes, and the authors of *The Federalist Papers* to the present ability of liberal societies to make policy for risk. Also, in chapter three, the Greek understanding of authority is a subject for exploration within the present context of risk's effects in modern society.

3. See, for example, Michael D. Kraft and Bruce B. Clary (1991), "Citizen Participa-

tion and the NIMBY Syndrome: Public Response to Radioactive Waste Disposal," *Western Political Quarterly* 44:299–328; Charles E. Davis (1986), "Public Involvement in Hazardous Waste Siting Decisions," *Polity* 19:296–304; Roger E. Kasperson (1986), "Six Propositions on Public Participation and Their Relevance to Risk Communication," *Risk Analysis* 6:275–81; Albert R. Matheny and Bruce A. Williams (1985), "Knowledge v. NIMBY: Assessing Florida's Strategy for Siting Hazardous Waste Disposal Facilities," *Policy Studies Journal* 14:70–80; and David Morell and Christopher Magorian (1982), *Siting Hazardous Waste Facilities: Local Opposition and the Myth of Preemption* (Cambridge, MA: Balinger).

4. See Anne L. Hiskes and Richard P, Hiskes (1986), *Science, Technology, and Policy Decisions* (Boulder, CO: Westview), among countless other works for some examples of citizens successfully stopping public works.

5. Nicholas Rescher (1983), *Risk* (Washington, DC: University Press of America), p. 5.

6. The idea of moral risks may seem somewhat odd, yet what is the new communitarian critique or the agenda of the religious right if not a call for our attention to the risks to morality today? Surely all would agree that some technologies such as television or reproductive techniques put some traditional ideas of morality at risk. Annette Baier perhaps goes too far, yet she performs an interesting philosophical juxtaposition of risk and morality by asserting that all of morality is "selecting which harms to worry about." See her "Poisoning the Wells," in Douglas MacLean, ed. (1986), *Values at Risk* (Lanham, MD: Rowman & Littlefield). See also Jean-Jacques Laffont (1995), "Regulation, Moral Hazard and Insurance of Environmental Risks," *Journal of Public Economics* 58:319–36.

7. See Israel Sheffler (1982), *Science and Subjectivity* (Indianapolis: Hackett); Karl Popper (1934), *The Logic of Scientific Discovery* (New York: Harper & Row); and Thomas Kuhn (1970), *The Structure of Scientific Revolutions*, 2nd edition (Chicago: University of Chicago Press).

8. See the following articles reprinted in May Brodbeck (1968), *Readings in the Philosophy of the Social Sciences* (New York: Macmillan), pp. 245–335. "Social Facts," by Emile Durkheim; "Holism Versus Individualism," by Ernest Gellner; "Methodological Individualism and Social Tendencies," by J. W. N. Watkins; "Methodological Individualism: Definition and Reduction," by May Brodbeck; "Levels of Explanation in History," by A. M. MacIver; and "The Individual and the Marxist Philosophy of History," by Laird Addis.

9. Richard S. Rudner (1966), *Philosophy of Social Science* (Englewood Cliffs, NJ: Prentice-Hall), p. 71.

10. Paul Slovic (1992), "Public Perceptions of Risk," *Risk Management* March, p. 57.

11. Ulrich Beck (1992), *Risk Society* (London: Sage). Original German edition published in 1986 as *Risikogesellschaft: Auf dem Weg in eine andere Moderne* (Frankfurt: Suhrkamp Verlag).

12. For a discussion of Beck's central concept of "reflexive modernity," or "self-monitoring" society, see Scott Lash (1993), "Reflexive Modernization: The Aesthetic Dimension," *Theory, Culture & Society* 10:1–23.

13. See, for instance, Baruch Fischhoff et. al. (1981), *Acceptable Risk* (Cambridge: Cambridge University Press), and Rescher (1983) as examples of theorists who acknowledge the two aspects of risk but avoid choosing between them. In the next section we engage several risk theorists who do not avoid the choice.

14. Roger E. Kasperson (1992), "The Social Amplification of Risk: Progress in Developing an Integrative Framework," in Sheldon Krimsky and Dominic Golding, eds., *Social Theories of Risk* (Westport, CT: Praeger), p. 155.

15. "The New Legal Structure of Risk Control," in Peter J. Burger, Jr., ed. (1990), *Risk* (Ann Arbor: University of Michigan Press), p. 214.

16. Langdon Winner (1977), *Autonomous Technology* (Cambridge, MA: MIT Press).

17. The linguistic prejudice against passive voice is clearly a part of liberal individualist approaches to the whole concept of agency. As students, our teachers taught us that passive voice is vague because it is difficult to point out who is responsible for the action described by the verb. That, of course, is the point: emergent phenomena *do* resist reduction to individual persons or behaviors and thus often are best described using passive voice. Doing so does not make them any less real.

18. For two interesting accounts of plagues as models of risk and response, see William H. Foege (1988), "Plagues: Perceptions of Risk and Social Responses," *Social Research* 55(3):331–42; and David A. J. Richards (1988), "Human Rights, Public Health, and the Idea of Moral Plague," *Social Research* 55(3):491–528.

19. Harvey M. Sapolsky (1986), *Consuming Fears* (New York: Basic), p. 3.

20. The noncommunal, even anticommunal, charge against liberal individualism is 200 years old, of course, and stated powerfully by Marx in *On the Jewish Question*, and de Tocqueville in *Democracy in America*. More recent sources include Robert Paul Wolff, (1968), *The Poverty of Liberalism* (Boston: Beacon), and Robert Bellah et al. (1985), *Habits of the Heart* (Berkeley: University of California Press). For a dissenting view see my *Community Without Coercion, Getting Along in the Minimal State* (1980) (Newark: University of Delaware Press).

21. Some technologies potentially blur the lines between individuals and even between species. Cloning, of course, presents these possibilities, only potential (so far) for humans. But in agricultural biotechnology, the recombination of genes *between* species has had wide impact and offers both new possibilities and risks for feeding the world. See Bette Hileman (1995), "Views Differ Sharply Over Benefits, Risks of Agricultural Biotechnology," *Chemical & Engineering News* August 21:8–17.

22. Mary Douglas and Aaron Wildavsky (1982) *Risk and Culture* (Berkeley: University of California Press). Other important contributions to cultural theories of risk include Michael Thompson (1980), "Aesthetics of Risk: Culture or Context?" in R. C. Schwing and W. A. Albers, eds., *Societal Risk Assessment: How Safe is Safe Enough?* (New York: Plenum); Steve Rayner (1990), *Risk in Cultural Perspective: Acting Under Uncertainty* (Norwell, MA: Kluwer); Niklas Luhmann (1993), *Risk: A Sociological Theory* (New York: Aldine DeGruyter); Sam Cole (1990), "Cultural Technological Futures," *Alternatives* 15:373–400; and Stephen H. Linder (1995), "Contending Discourses in the Electric and Magnetic Fields Controversy: The Social Construction of EMF Risk as a Public Problem," *Policy Sciences* 28:209–30.

23. Douglas and Wildavsky, p. 80.

24. Chauncey Starr (1969), "Social Benefit versus Technological Risk," *Science* 165: 1232–38.

25. Ibid., p. 1232.

26. See S. Lichtenstein, et.al. (1978), "Judged Frequency of Lethal Events, "*Journal of Experimental Psychology: Human Learning and Memory* 4:551–78.

27. See Fischhoff et al. (1981).

28. Ibid., p. 81.

29. Robert E. Goodin (1982), *Political Theory and Public Policy* (Chicago: University of Chicago Press), chapter 8. Paul Slovic also points out that the revealed preference model "enshrines current economic and social arrangements, assuming that accepted risks are acceptable risks." In "Perception of Risk: Reflections on the Psychometric Paradigm," in Krimsky and Golding, p. 118.

30. Executive Order 12898; quoted in Michael E. Kraft and Denise Scherberle (1995), "Environmental Justice and the Allocation of Risk: The Case of Lead and Public Health," *Policy Studies Journal* 23:113.

31. See especially B. Bryant and P. Mohai, eds. (1992), *Race and the Incidence of Environmental Hazards: A Time for Discourse* (Boulder, CO: Westview); R. D. Bullard (1994), *Dumping in Dixie: Race, Class, and Environmental Decisionmaking* (Boulder, CO: Westview). For a review of recent literature and current disputes, see Christopher J. Paterson and Richard N. L. Andrews (1995), "Procedural and Substantive Fairness Risk Decisions: Comparative Risk Assessment Procedures, *Policy Studies Journal* 23:85–95; Robert R. Higgins (1993), "Race & Environmental Equity: An Overview of the Environmental Justice Issue in the Policy Process," *Polity* 26:281–300; and Kraft and Scheberle (1995).

32. See Nancy Tuana, ed. (1989), *Feminism and Science* (Bloomington: Indiana University Press; Michel Foucault (1977), *Discipline and Punish* (New York: Pantheon); and Jacques Derrida (1992), *Acts of Literature* (New York: Routledge).

33. Sheldon Krimsky (1992), "The Role of Theory in Risk Studies," in Krimsky and Golding (1992), p. 18.

34. See K. S. Shrader-Frechette (1991), *Risk and Rationality* (Berkeley: University of California Press), chapter 3. For a review and discussion of these issues, see William Leiss, (1992), Book Review Essay: "Assessing and Managing Risks," *Policy Sciences* 25:341–49.

35. Dorothy Nelkin (1982), "Blunders in the Business of Risk," *Nature* 298:775–76.

36. Michael Thompson (1986), "To Hell with the Turkeys! A Diatribe Directed at the Pernicious Trepidity of the Current Intellectual Debate on Risk," in MacLean (1986), p. 114.

37. Maclean (1986), p. 11.

38. Shrader-Frechette (1991), p. 31.

39. Ibid. A similar accusation could be made against Thompson in "To Hell with the Turkeys!" (1986). His ideological bias in favor of British risk policy and against policy in the US is evident throughout his article, although he denies it when he argues that "[c]ultural analysis (unlike systems analysis, with its in-built bias of toadying to power) is, quite properly, silent on whether regimes should be upheld or overthrown" (p. 124).

40. Shrader-Frechette, p. 30.

41. Luhmann (1993), chapter one.

42. Sheila Jasanoff (1993), "Bridging the Two Cultures of Risk Analysis," *Risk Analysis* 13(2):126.

43. See Dorothy Nelkin (1979), *Controversy: Politics of Technical Decisions* (Beverly Hills, CA: Sage).

44. Hiskes and Hiskes (1986), pp. 1–4.

45. Beck (1992), p. 183.

46. Ingmar Palmlund (1992), "Social Drama and Risk Evaluation," in Krimsky and Golding, pp. 197–212.

47. Brian Wynne (1980), "Technology, Risk, and Participation: On the Social Treatment of Uncertainty," in J. Conrad, ed. (1990), *Society, Technology, and Risk* (New York: Academic Press), pp. 167–202.

48. Brian Wynne (1992), "Risk and Social Learning: Reification to Engagement," in Krimsky and Golding, p. 281.

49. William Leiss and Christina Chociolko (1994), *Risk and Responsibility* (Montreal: McGill-Queen's University Press), p. 6.

50. Sheila Jasanoff, paper presented at the Symposium on Managing the Problem of Industrial Hazards: The International Policy Issues; National Academy of Sciences, Washington, DC, February 27, 1989. Quoted in M. Granger Morgan, Baruch Fischhoff, et al. (1992), "Communicating Risk to the Public," *Environmental Science and Technology* 26:2048–56, p. 2048.

51. Morgan et al. (1992), p. 2055.

52. Kuhn himself disagrees that he must conclude this, but some who interpret him so

include Alan C. Isaac (1969), *Scope and Methods of Political Science* (Homewood, IL: Dorsey), and David Ricci (1984), *The Tragedy of Political Science* (New Haven: Yale). Both Isaac and Ricci are, unsurprisingly, political scientists.

53. I trust it is not necessary to cite everyone who included these two elements into his or her definitions of politics. Some notable exceptions, of course, are most premodern theorists, beginning with Plato, and also some contemporary inheritors of Rousseau's belief that politics could be about consensus rather than conflict. I do not want to belittle such conceptions, but I do think that both the classical view and Rousseau's resuscitation of it require the exercise of power (cf. Plato's Myth of the Metals and Rousseau's Legislator and the General Will, and both's acceptance of censorship) to instill the necessary belief in the possibility of consensus. For detailed treatments of the concept of power, see, for example, Felix E. Oppenheim (1981), *Political Concepts: A Reconstruction* (Chicago: University of Chicago); William E. Connolly (1974), *The Terms of Political Discourse* (Lexington, MA: D. C. Heath); and Brian Barry, ed. (1977), *Power and Political Theory* (London: John Wiley & Sons).

54. I don't really want to be sidetracked into a discussion over what exactly is the focal point of evolution. For some different points of view, see Darwin's (1988) *Origin of Species* (New York: NYU Press); Richard Dawkins (1976), *The Selfish Gene* (New York: Oxford); and Jacob Bronowski (1973), *The Ascent of Man* (Boston: Little, Brown).

55. Oppenheim (1981) (chs. 2, 3) sees an important distinction in terms of the risks posed to others between having and actually exercising power. Unfortunately, it is a distinction that tends to dissolve for the most risky of technologies.

56. Peter Bachrach and Morton S. Baratz (1962), "Two Faces of Power," *American Political Science Review* 56:947–52.

57. Ibid., p. 950.

58. 381 US 479 (1965).

59. Carole Pateman (1970), *Participation and Democratic Theory* (Cambridge: Cambridge University Press), pp. 42, 43.

60. See, for example, Benjamin Barber (1984), *Strong Democracy* (Berkeley: University of California Press); Peter Bachrach (1980), *The Theory of Democratic Elitism* (Washington, DC: University Press of America); John T. Rourke, Richard P. Hiskes, and Cyrus Ernesto Zirakzadeh (1991), *Direct Democracy and International Politics* (Boulder, CO: Lynne Rienner); and my (1992) "The Democracy of Risk," *Industrial Crisis Quarterly* 6(4):259–78.

61. Jurgen Habermas (1984), *The Theory of Communicative Action* (Boston: Beacon).

62. Michel Foucault (1976), *The History of Sexuality*, Vol. 1 (New York: Pantheon), p. 143, quoted in Paul Rabinow, ed. (1984), *The Foucault Reader* (New York: Pantheon), p. 17.

63. Ibid.

64. Ibid, quoted in Rabinow, ed. (1984), p. 266.

65. Theodore J. Lowi (1990), "Risks and Rights in the History of American Governments," in Peter J. Burger Jr., ed. (1990), *Risk* (Ann Arbor: University of Michigan Press), p. 30.

66. See Paul Slovic, James H. Flynn, and Mark Layman (1991), "Perceived Risk, Trust, and the Politics of Nuclear Waste," *Science* 254:1603–7.

67. Lowi (1990), p. 19.

68. For a good discussion of siting politics in general, see Gerald Jacob (1990), *Site Unseen: The Politics of Siting a Nuclear Waste Repository* (Pittsburgh: University of Pittsburgh Press).

69. See Rourke, Hiskes, and Zirakzadeh (1991), chapter 6, for a discussion of the freeze movement.

70. It is not clear, however, what the connection is between the numbers of citizens roused to action to fight a political battle (either at the national, state, or local levels) involving risks—such as siting disputes—and the degree of participation in traditional electoral politics. This is a fertile area for research by political scientists interested in the general topic of participation. Exploring the politics of risks at least exposes, however, the inadequacy of some current definitions of what counts as political participation within the political science literature.

71. Rabinow (1984), p. 13.

72. See Nelkin (1979), for case study examples of several such disputes. See, also, Brian Wynne, ed. (1987), *Risk Assessment and Hazardous Waste Management: Implementation and the Dialectics of Credibility* (London: Springer-Verlag), and Steve Rayner (1992), "Cultural Theory and Risk Analysis," in Krimsky and Golding.

2. *Risk, Consent, and Communal Identity*

1. Chauncey Starr (1969), "Social Benefit versus Technological Risk," *Science* 165:1232–38, p. 38.

2. Mary Douglas and Aaron Wildavsky (1982), *Risk and Culture* (Berkeley: University of California Press), p. 5. Emphases theirs.

3. Baruch Fischhoff et al. (1981), *Acceptable Risks* (Cambridge: Cambridge University Press), p. xii.

4. For a discussion of the role of consent assumptions regarding the siting of radioactive waste disposal dumps, see Gerald Jacob (1990), *Site Unseen: The Politics of Siting a Nuclear Repository* (Pittsburgh: University of Pittsburgh Press).

5. Court cases include *Salgo v. Leland Stanford Jr. University Board of Trustees*, 317 P. 2d 170, 181 (Cal. Dist. CT. App. 1957); *Natanson v. Kline*, 350 P. 2d 1093, 1104 (Kansas, 1960); *Canterbury v. Spence*, 464 F. 2d 772, 780 (DC. Cir. 1972); and *Scott v. Bradford*, 606 P. 2d 554, 556 (Okla. 1979). Useful scholarly discussions include Ruth Faden and Thomas Beauchamp (1986), *A History and Theory of Informed Consent* (New York: Oxford); Paul S. Appelbaum, Charles W. Lidz, and Alan Meisel (1987), *Informed Consent: Legal Theory and Clinical Practice* (New York: Oxford); Gerald Dworkin, ed. (1988), *The Theory and Practice of Autonomy* (Cambridge: Cambridge University Press); and Martin Gunderson (1990), "Justifying a Principle of Informed Consent: A Case Study in Autonomy-Based Ethics," *Public Affairs Quarterly* 4:249–65; July.

6. See, for instance, K. S. Shrader-Frechette (1993), "Consent and Nuclear Waste Disposal," *Public Affairs Quarterly* 7:363–77, for her use of professional ethics models in evaluating the siting of the Yucca Mountain radioactive waste dump. Recent accounts of professional ethics and their reliance on consent to risks include T. L. Beauchamp and J. F. Childress (1989), *Principles of Biomedical Ethics* (New York: Oxford); M. Bayles (1981), *Professional Ethics* (Belmont, CA: Wadsworth); and M. Curry and Larry May (1984), *Professional Responsibility for Harmful Actions* (Dubuque, IA: Kendal/Hunt).

7. See Fischhoff et al. (1981), especially chapter 4.

8. Douglas MacLean (1986), "Risk and Consent: Philosophical Issues for Centralized Decisions," in Douglas MacLean, ed., *Values at Risk*, Lanham, MD: Rowman & Littlefield, Inc., p. 19.

9. Ibid., p. 20.

10. Starr (1969), p. 1232.

11. Fischhoff et al., (1969), p. 81.

12. Kristin Shrader-Frechette (1985), *Risk Analysis and Scientific Method* (Dordrecht: D. Reidel), p. 37.

13. Robert Goodin (1982), *Political Theory and Public Policy* (Chicago: University of Chicago Press), pp. 147ff.

14. The current discussions concerning concentrations of lead poisoning occurrences in urban ghettoes is a good example of how a recent identification of a risk might lead to further progress for blacks. Of course, the examples of the civil rights and women's movements also hint at how the distribution of modern risks can lead to new manifestations of identity politics. This possibility is explored at length in chapter six.

15. The term "snapshot" to indicate the short-term relevance of any poll during a campaign is taken from discussions with G. Donald Ferree, Director of the Roper Poll.

16. An interesting application of such reasoning concerning surveys and consent is the recent "National Issues Convention" (NIC) that took place in Austin, Texas, in winter 1996. Having identified a random sample of citizens by means of a typical national poll, these citizens were invited to Austin to participate in a weekend of "deliberation" about issues and candidates in the 1996 presidential election. Its organizer, Professor James Fishkin of the University of Texas, then polled the participants at the end of the sessions and concluded that their responses to this "deliberative poll" represented how American citizens as a whole would have voted if they had similar access to such information concerning issues and actual interviews with the candidates. His reasoning is clearly fallacious, even given the laudatory premises and results of the convention. For an excellent discussion of how to interpret the NIC, see Catherine Flavin and Regina Dougherty (1996), "Science and Citizenship at the NIC," *The Public Perspective* 7:46–9; April/May.

17. Shrader-Frechette (1985), p. 44.

18. See D. M. Settle and C. C. Patterson (1980), "Lead in Albacore: Guide to Lead Pollution in Americans." *Science* 207:1167–76.

19. *Code of Federal Regulations*, 10, part 20, US Government Printing Office, Washington, DC., 1978, p. 189. Cited in Shrader-Frechette (1985), p. 45.

20. Goodin (1982), pp. 193–4.

21. See Shrader-Frechette (1985), pp. 46–7, and Fischhoff et al. (1981), pp. 86–8.

22. It should be clear that in pointing out this essential similarity of all NIMBY claims, I am not thereby oversimplifying their complexity or the extent to which different NIMBY disputes might differ from each other. Some such claims might border on selfishness; some might genuinely seek to further the wider social good. Some might be based on Ludditism; some might demonstrate considerable scientific sophistication. It is difficult to generalize across all such claims, except to state the almost tautological truth that, since they are claims—in many cases public disputes—NIMBY episodes are *actions*. As acts, they focus on a question of whether previous lack of action has resulted in consent; thus, they all deny the legitimacy of tacit consent, based as it is on just such lack of action.

23. See note 3.

24. Carole Pateman (1979), *The Problem of Political Obligation* (New York: John Wiley & Sons), p. 88.

25. John Locke (1952), *The Second Treatise of Government*, Thomas P. Peardon, ed. (Indianapolis: Bobbs-Merrill), section 119, p. 68.

26. As an early state of nature theory, Locke's of course assumes that the standards nature provides should be a guide for politics. This reliance on "the book of nature" becomes a powerful tool for argument within seventeenth- and eighteenth-century liberal thought. The natural standards model of risk assessment is, in that sense as well as others, a classically liberal response to risk.

27. Locke (1952), p. 20.

28. Ibid., p. 29.

29. Ibid., p. 71.

30. Two examples among many are Richard Ashcraft, who believes it possible for Locke to argue that "the invention and use of money can be shown to be compatible with the advancement of the common good of mankind" and, therefore, "fall under the authoritative endorsement of the natural law" and C. B. MacPherson, who insists that for Locke "the introduction of money . . . has invalidated the natural provision that everyone should have as much as he could make use of." See Ashcraft (1986), *Revolutionary Politics & Locke's Two Treatises of Government* (Princeton: Princeton University Press), p. 274, and MacPherson (1962), *The Political Theory of Possessive Individualism* (Oxford: Oxford University Press), pp. 203–4.

31. Craig L. Carr (1990), "Tacit Consent," *Public Affairs Quarterly* 4:335–45; October. Carr argues that Locke's use of tacit consent both to presume an already existing moral community and to explicate how it can be legitimately changed without violating the requirements of individual liberty is firmly grounded in sixteenth- and seventeenth-century usage.

32. David-Hillel Ruben (1982), "Tacit Promising," *Ethics* 83:71–9, pp. 77–8.

33. I take this to be the crux of C. B. MacPherson's (1962, p. 210) argument about obligation in Locke's theory. By arguing that there are "two levels of consent in Locke's theory," the first, given (tacitly) to money, effectively outweighs the second (to government), which, for MacPherson, is necessary only to enforce the first (p. 210).

34. Rousseau states in the *Discourse on the Origin of Inequality* that the first permanent families were the result of the random coming together (the "novel situation") of male and female members of the species beneath the same tree for more than momentary, usually sexual, purposes. See G. D. H. Cole, ed. (1950), *Rousseau: The Social Contract and Discourses* (New York: Everyman), p. 239.

35. Will Kymlicka (1990), *Contemporary Political Philosophy: An Introduction* (Oxford: Clarendon Press), p. 59. He continues, [social contract arguments] "ask us to imagine a state of nature before there is any political authority. Each person is on their own, in the sense that there is no higher authority with the power to command their obedience, or with the responsibility for protecting their interests or possessions. The question is, what kind of contract would such individuals, in the state of nature, agree to concerning the establishing of a political authority which would have these powers and responsibilities?"

36. Ronald Dworkin (1977), *Taking Rights Seriously* (London: Duckworth), p. 151, quoted in Kymlicka (1990), p. 59.

37. See Carr (1990); see also Craig L. Carr (1988), "Coercion and Freedom," *American Philosophical Quarterly* 25:59–67. For a different interpretation of social contract agreements that construe them not as contracts but "joint agreements," see Margaret Gilbert (1993), "Agreements, Coercion, and Obligation," *Ethics* 103:679–706. As such agreements or contracts affect future generations, see Matthew Hanser (1990), "Harming Future People," *Philosophy & Public Affairs* 19:47–70.

38. Pateman (1979), p. 8. Her position is chiefly a critique of Michael Walzer (1970), *Obligations: Essays on Disobedience, War, and Citizenship* (New York: Clarion), but is equally applicable to the later approaches of Rawls and Nozick. Kymlicka (1990) and Michael J. Sandel (1982), *Liberalism and the Limits of Justice* (Cambridge: Cambridge University Press) apply a similar argument—that leads to different proposals for liberal society—to Rawls and Nozick.

39. Carr (1990), p. 337, makes this point.

40. Locke (1952, p. 79) defines the lack of clarity by stating that three things are missing in state of nature: a singular, authoritative interpretation of the natural law, an impartial judge, and enforcement power.

41. A well-known example of this truth within social contract theory is Rousseau's understanding of what is often called the "Golden Age" of human political development. During this period on the evolutionary track to the modern state, government does not yet exist, yet actual communities of families are based on relationships of duty and obligation, and even rudimentary understandings of natural rights keep social life relatively peaceful and free of the debilitating dependence that accelerates both inequality and the rise of the state. See Rousseau in Cole (1950); Judith Shklar (1969), *Men and Citizens* (London: Cambridge University Press).

42. See Brian Barry (1968), "Warrender and His Critics," Philosophy 43:117–37.

43. See Hume's discussion of "sympathy" as an element of moral obligation in the *Treatise*, especially Part 3. Henry D. Aiken, ed. (1972), *Hume's Moral and Political Philosophy* (New York: Hafner), pp. 131–69.

44. See Benjamin Barber (1984), *Strong Democracy* (Berkeley: University of California Press). For an interpretation of what "acts of democracy" might entail obligations, see my (1996) "Acts of Democracy: Reconceptualizing Politics, Participation, and Competence," *The Public Perspective* 7:40–4, June/July.

45. Classical critics include Comte, Marx, and anarchists like Bakunin; contemporary ones include Robert Paul Wolffe (1968), *The Poverty of Liberalism* (Boston: Beacon); Michael J. Sandel (1982), *Liberalism and the Limits of Justice* (Cambridge: Cambridge University Press); Alistair MacIntyre (1981), *After Virtue* (Notre Dame, IN: Notre Dame University Press); and Charles Taylor, "Cross Purposes: The Liberal-Communitarian Debate," in Nancy L. Rosenblum (1989), *Liberalism and the Moral Life* (Cambridge, MA: Harvard University Press), pp. 159–82. Contemporary defenders of liberalism's communitarian possibilities include my own (1982) *Community Without Coercion: Getting Along in the Minimal State* (Newark: University of Delaware Press); Will Kymlicka (1989), *Liberalism, Community, and Culture* (Oxford: Clarendon); and William A. Galston (1991), *Liberal Purposes* (Cambridge: Cambridge University Press).

46. Sandel (1982), pp. 106–7.

47. Sandel (1982), p. 62.

48. See William A. Galston (1982), "Defending Liberalism," *American Political Science Review* 76:621–9. Galston elaborates his full argument in Galston (1991).

49. Galston (1991), p. 3.

50. Galston (1982), p. 621. The works he refers to are Rawls (1971), *A Theory of Justice* (Cambridge, MA: Harvard University Press); Dworkin (1977), *Taking Rights Seriously* (Cambridge, MA: Harvard University Press); William Ackerman (1980), *Social Justice in the Liberal State* (New Haven: Yale University Press).

51. Galston (1982), p. 8.

52. In his earlier version of the argument (Galston, 1982, p. 626, n. 2), Galston even includes libertarian Robert Nozick in his list of misunderstood liberal communitarians. Though I admit to the same gambit (Hiskes, 1982, ch. 4), Galston drops Nozick from the latter (1991), more fully developed argument.

53. Locke (1952), p. 11.

54. Roberto M. Unger (1986), *The Critical Legal Studies Movement* (Cambridge, MA: Harvard University Press).

55. See Christopher J. Berry (1993), "Shared Understanding and Democratic Way of Life," in John W. Chapman and Ian Shapiro (1993), *NOMOS XXXV: Democratic Community* (New York: New York University Press), pp. 67–87, p. 79.

56. Within eighteenth-century contract theory, only Tom Paine and Thomas Jefferson can be said to recognize the necessity of expressed consent and of the guarantee that it need

not be given once and for all. Paine argues for a constitutional convention every generation to give every citizen a chance to render (or withhold) expressed consent; Jefferson accepts "a little revolution" every now and then, for the same purpose.

57. Amy Gutmann (1993), "The Disharmony of Democracy," in Chapman and Shapiro (1993), pp. 126–60.

58. William E. Connolly (1991), *Identity\Difference: Democratic Negotiations of Political Paradox* (Ithaca, NY: Cornell University Press), p. 64.

59. Connolly (1991), p. 65.

3. *Risk and the Authority of Ends*

1. As Joseph Raz states, "consent by itself, actual consent to a social contract, cannot solve the problem. But a theory of participatory government in a society in which conditions exist that enable each to see his own well-being as tied up with the prosperity of others can do so. The development of such a theory has been the aim of political philosophy since Rousseau. Its achievement is eagerly awaited." Raz, ed. (1990), *Authority* (New York: New York University Press), p. 17. Such participation would constitute actual consent on an ongoing basis, a point similar to that made by Amy Gutmann and referred to in the previous chapter.

2. William E. Connolly (1987), "Modern Authority and Ambiguity," in J. Roland Pennock and John W. Chapman (1987) *NOMOS XXIX, Authority Revisited* (New York: New York University Press), p. 10.

3. Sheldon Wolin (1960), *Politics and Vision* (New York: Little, Brown), p. 301.

4. The term is from the book of the same name by Thomas M. Dietz and Robert W. Rycroft (1987) (New York: Russell Sage). Dietz and Rycroft use the term broadly to refer to a large group of "policy elites," inhabiting governmental agencies, corporations, think tanks, and academic institutions. Their valuable work is largely descriptive, although it raises the interesting question concerning whether these powerful new group of leaders constitutes an elite, a class, an establishment, or a community.

5. Cited in Roger E. Kasperson (1986), " Six Propositions on Public Participation and Their Relevance for Risk Communication." *Risk Analysis* 6:275–81, p. 277. For discussions concerning the loss of trust specifically in the DOE and the NRC, see Paul Slovic, James H. Flynn, and Mark Layman (1991), "Perceived Risk, Trust, and the Politics of Nuclear Waste," *Science* 254:1603–7, and K. David Pijawka and Alvin H. Mushkatel (1991), "Public Opposition to the Siting of the High-Level Nuclear Waste Repository: The Importance of Trust," *Policy Studies Review* 10(4):180–94.

6. As Joseph Raz (1987) states, "[C]onsenting to be ruled by someone expresses confidence in that person's ability to rule well." Consent without such a presumption would be a hollow right indeed, as Raz's argument effectively demonstrates. "Government by Consent," in Pennock and Chapman, p. 91.

7. See Joseph Raz (1990), p. 122ff; also, Raz (1987), p. 91.

8. Don K. Price (1985), *America's Unwritten Constitution: Science, Religion, and Political Responsibility* (Cambridge, MA: Harvard University Press), p. 5.

9. Price (1985), p. 129.

10. The closest contemporary statement to "authority is legitimate power" perhaps is George E.G. Catlin's (1958) phrase, "authority is power exercised in accordance with convention." In Carl Friedrich, ed. (1958), *NOMOS I: Authority* (Cambridge, MA: Harvard University Press), p. 29. Not as far distant from Catlin's definition as it first reads is Russell Hardin's (1987) conclusion that "we may not wish to say that might is right, but only that opposition to it is wrong." In Pennock and Chapman, eds. (1987), p. 216.

11. Connolly (1987).

12. See Raz (1990), Introduction, p. 12.

13. Hannah Arendt (1961), "What is Authority?" in *Between Past and Future* (New York: Viking), p. 104.

14. Hannah Arendt (1958), "What Was Authority?" in Friedrich, p. 89.

15. Arendt (1961), p. 90.

16. Arendt (1961), p. 95. John Schaar (1981) echoes Arendt's belief that authority in the true, classical sense has disappeared. He writes, "I believe that genuine authority is all but lost to us today, and perhaps we have lost even the concept of authority, so that we cannot know what honorable obedience consists in. This is no small loss." *Legitimacy in the Modern State* (New Brunswick, NJ: Transaction), p.2.

17. Arendt (1961), p. 122.

18. Arendt (1961), p. 123.

19. Arendt's views on authority—indeed on several topics—are enjoying a renewed interest in political theory, particularly among feminist thinkers (a development that likely would have surprised Arendt). For an interesting application of feminist "care" theory to the concept of authority, resulting in an Arendtian view of authority as "augmentation"—in this instance of "compassion"—see Kathleen B. Jones (1987), "On Authority: Or, Why Women are Not Entitled to Speak," in Pennock and Chapman. See also Jones's expanded argument in her (1993), *Compassionate Authority: Democracy and the Representation of Women* (New York: Routledge).

20. Terry Ball (1987) uses the term "epistemocratic" authority as referring to authority "tied to claims of technical expertise or specialized knowledge." Thus, it is another term for "an" authority, to be distinguished from "in" authority. He, like Arendt, identifies "the *locus classicus* of the epistemocratic vision [as] Plato's *Republic*." In Pennock and Chapman, pp. 39, 49. An earlier, similar characterization of "an" authority is in Richard T. DeGeorge, "The Nature and Function of Epistemic Authority," in R. Baine Harris (1976), *Authority: A Philosophical Analysis* (Tuscaloosa: University of Alabama Press).

21. Richard B. Friedman (1973), "On the Concept of Authority in Political Philosophy," in Richard E. Flathman (1973), *Concepts in Social and Political Philosophy* (New York: Macmillan), pp. 121–46, p. 123.

22. Alasdair MacIntyre (1973), "Secularization and Moral Change," in Flathman (1973), pp. 163–67, p. 165.

23. Henri Bergson (1935), *The Two Sources of Morality and Religion*, (Garden City, NY: Doubleday Archor), pp. 9-26. Cited by Friedman (1973), in Flathman (1973).

24. Elizabeth Anscombe (1973), "Authority in Morals," in Flathman (1973), pp. 157–63, p. 158.

25. Friedman (1973), p. 136.

26. Robert Paul Wolff (1970) (New York: Harper); see also Raz (1987, 1990); Nancy Rosenblum (1987), "Studying Authority: Keeping Pluralism in Mind," in Pennock and Chapman; Richard E. Flathman (1980), *The Practice of Political Authority* (Chicago: University of Chicago Press); and Mark E. Warren (1996), "Deliberative Authority and Democracy," *American Political Science Review* 90:46–60. I return to Warren's argument later in this chapter. He goes so far in denying that "an" authority requires the suspension of independent judgment by subjects as to require the existence of "attentive nonexperts" to validate the ("an") authority of experts (p. 56).

27. An interesting new take on Machiavelli's meaning and purpose in *The Prince* seeks explicitly to tie his political agenda to his contempt for clerical power. See John T. Scott and Vickie B. Sullivan (1994), "Patricide and the Plot of *The Prince*: Cesare Borgia and Machiavelli's Italy," *American Political Science Review* 88 (4):887–900.

28. Machiavelli (1952), *The Prince* (New York: New American Library), chapter 9, p. 63.

29. See Thomas L. Pangle (1992), *The Ennobling of Democracy* (Baltimore: Johns Hopkins University Press), p. 134. Emphasis Pangle's.

30. Price (1983), p. 9.

31. Price (1983), p. 93.

32. James A. Thurber (1994), "'Not in My Back Yard': High-Level Nuclear Waste Policy," in Robert S. Gilmour and Alexis A. Halley, *Who Makes Public Policy: the Struggle for Control Between Congress and the Executive* (Chatham, NJ: Chatham House), p. 71.

33. George Kateb (1992), *The Inner Ocean* (Ithaca, NY: Cornell University Press,) pp. 37–38. See also Warren (1996), p. 47.

34. Kateb (1992), p. 39.

35. Kateb (1992), p. 40.

36. Pijawka and Mushkatel (1991), p. 180. See also T. R. Mitchel and W. G. Scott (1987), "Leadership Failures, the Distrusting Public, and Prospect of the Administrative State," *Public Administration Review* 47:445–52; cited in Pijawka and Mushkatel.

37. Nancy L. Rosenblum (1987), "Studying Authority: Keeping Pluralism in Mind," in Pennock and Chapman, pp. 111–12.

38. Price (1993), p. 4.

39. Richard S. Peters, in Flathman (1973), p. 155.

40. Several histories of the relationship between government and science exist. Don K. Price's (1965) *The Scientific Estate* (Cambridge, MA: Harvard University Press) is still one of the best. See also Harvey Brooks (1968), *The Government of Science*, (Cambridge: MIT), and Anne L. and Richard P. Hiskes 1986, *Science, Technology, and Policy Decisions* (Boulder, CO: Westview), especially chapter 2.

41. Price (1983), p. 142.

42. See Hiskes and Hiskes (1986).

43. Quoted in Dorothy Nelkin (1979), *Controversy: Politics of Technical Decisions* (Beverly Hills, CA: Sage), pp. 16–7.

44. Warren (1996), p. 51.

45. See Slovic, Flynn, and Layman (1991).

46. "In A-Plants' Backyard in Connecticut, Trust Ebbs," *New York Times*, Wednesday, April 3, 1996, pp. A1, B4.

47. The term "attentive nonexperts" is Warren's (1996).

48. George L. Priest, "The New Legal Structure of Risk Control," in Edward J. Burger, Jr., ed. (1990), *Risk* (Ann Arbor: University of Michigan Press), p. 207.

49. Peter W. Huber, "Pathological Science in Court," in Burger, Jr., pp. 97–118, p. 98.

50. Huber (1990), p. 100.

51. Huber (1990), p. 101.

52. Huber (1990), p. 113.

53. Quoted in "Genetic Questions are Sending Judges Back to Classroom," *New York Times*, July 9, 1996, pp. C1, C9.

54. Ibid.

55. Priest (1990), p. 209.

56. Priest (1990), p. 214.

57. Priest (1990), p. 215.

58. Douglas MacLean (1986), *Values at Risk* (Lanham, MD: Rowman & Littlefield), pp. 8–9.

59. See Dorothy Nelkin (1984), "Science and Technology Policy and the Democratic

Process," in James C. Petersen, ed., (9184) *Citizen Participation in Science Policy* (Amherst: University of Massachusetts Press), p. 26.

60. Kim Lane Scheppele and Karol Edward Soltan (1987), "The Authority of Alternatives," in Pennock and Chapman, pp. 169–200, p. 175.

61. Ibid., p. 171.

62. Flathman (1980), pp. 96–97.

4. *Risk, Self-Interest, and the New Technology of Rights*

1. Cass Sunstein (1990), "Constitutional Politics and the Conservative Court," *The American Prospect* 1:51–61, p. 60.

2. The term is Mary Ann Glendon's, who uses it as the title of her recent book (1991) (New York: Free Press).

3. Jeremy Bentham, "Anarchical Fallacies," in Jeremy Waldron (1987), *Nonsense Upon Stilts: Bentham, Burke and Marx on the Rights of Man* (London: Methuen); p. 53.

4. L. W. Sumner (1987), *The Moral Foundation of Rights* (Oxford, England: Clarendon Press), p. 13.

5. See Morton E. Winston (1989), *The Philosophy of Human Rights* (Belmont, CA: Wadsworth), p. 8ff, for a discussion of the differences between moral and legal and civil rights.

6. Judith Jarvis Thomson (1990), *The Realm of Rights* (Cambridge, MA: Harvard University Press), pp. 70–76.

7. Waldron, p. 163.

8. Thomson (1990) uses the same type of argument in arguing against the "two species" thesis and replacing it with a sense that rights can have two (or several) different sources. She states that "it is no doubt hard to imagine a legal system that does not assign to anyone living under it a right to not be murdered; perhaps that is an unimaginable legal system. But it is unfortunately all too easy to imagine a legal system that does not assign to (as it might be) Jews or blacks a right to not be murdered" (p. 74).

9. Diana T. Meyers (1985), *Inalienable Rights: A Defense* (New York: Columbia University Press). Joel Feinberg's general argument is similar to Meyers's. He argues that rights are necessary in order to stipulate duties and to achieve any notion of moral worth—an idea necessary to all social life and absent only in his hypothetical (and unappealing) "Nowheresville." See Feinberg (1980), *Rights, Justice, and the Bounds of Liberty: Essays in Social Philosophy* (Princeton, NJ: Princeton University Press).

10. Joseph Raz, "On the Nature of Rights," in Winston (1989), pp. 44–60, p. 52.

11. Raz, p. 52.

12. Sumner (1987), chapter 4.

13. Raz, p. 50.

14. Meyers, pp. 34–35.

15. Richard Flathman (1976), *The Practice of Rights* (Cambridge: Cambridge University Press), p. 14ff.

16. Flathman, p. 185.

17. Feinberg (1980), p. 151.

18. See Raz (1989), pp. 55–57.

19. It is not my intention here to engage in the dispute over whether Mill is indeed a rule utilitarian who accepts rights as the products of such rules, or even whether rule utilitarianism makes sense as moral theory. All I will point out here is that the language of rights is indeed present in Mill's work, and, J. J. C. Smart's disdain notwithstanding, rule utilitari-

anism has been forwarded by moral theorists as a consistent moral philosophy, and those theorists typically take Mill as inspiration. See Samuel Gorovitz (1971), *Mill: Utilitarianism* (Indianapolis: Bobbs-Merrill), especially the articles by Smart (pp. 195–204), Rawls (pp. 175–94), and Diggs (pp. 306–23).

20. Ronald Dworkin (1977), *Taking Rights Seriously* (Cambridge, MA: Harvard University Press).

21. This is different than saying that rights do not account for all of the moral domain. Both Sumner and Thomson insist that the latter is true—that rights do not exhaust the moral domain. I agree, and the argument that follows obviously presumes that there are other elements to morality—both in terms of claims and duties—than those included in "rights-as-antes."

22. Simone Chambers (1993), "Talking About Rights: Discourse Ethics and the Protection of Rights," *The Journal of Political Philosophy* 1 (3):229–49, pp. 230, 233.

23. Chambers (1993), p. 238.

24. Robert E. McGinn (1994), "Technology, Demography, and the Anachronism of Traditional Rights," *Journal of Applied Philosophy* 11:57–70.

25. Ibid., p. 63.

26. Ibid., p. 66.

27. Meyers, pp. 53–54 and Feinberg, p. 199. Feinberg calls the right a "claim-right" following Wesley Hohfeld's influential formulation of the four kinds of rights. Briefly, a claim-right entitles one to an obligation owed him or her by someone else. The other three types of rights for Hohfeld are liberties, powers, and immunities. See Wesley Newcomb Hohfeld (1919, 1964), *Fundamental Legal Conceptions* (New Haven, CT: Yale University Press).

28. See, for instance, Judith Jarvis Thomson (1986), *Rights, Restitution, & Risk* (Cambridge, MA: Harvard University Press), especially chs. 2, 10, and Virginia Held (1984), *Rights and Goods* (Chicago: University of Chicago Press), especially chs. 11, 12.

29. Hugo Bedau (1968), "The Right to Life," *The Monist* 52:567, quoted in Feinberg, p. 222.

30. Thomson (1990), pp. 55–56.

31. My argument supporting this conclusion is in Hiskes and Hiskes (1986), *Science, Technology, and Policy Decisions* (Boulder, CO: Westview), chapter 8. See also Robert Blank (1981), *The Political Implications of Human Genetic Technology* (Boulder, CO: Westview). An interesting dispute concerning these issues surrounding the rights to life and to abortion erupted recently in London over the prospective disposal of 3,300 frozen human embryos no longer wanted or claimed by biological parents. *The New York Times* (September 1, 1996, pp. A1, A4) reported widespread outrage stimulated by the prospect of the first enforcement of a law requiring sperm banks to so dispose of the embryos. Many citizens and religious leaders proclaimed that the rights to life of the 4-celled embryos would be violated by such an act. At the very least, Cardinal Hume of England argued, the embryos should be given a proper burial. Although it probably is technically wrong to call disposal of such cryogenically preserved embryos actual abortions, since they were not (ever) in a womb, it is telling that the dispute erupted as if acts of abortion were actually going to occur.

32. I fully recognize that a view of rights as "antes" effectively excludes fetuses from being viewed as person with such rights. Thus, the rights-as-antes view is fully consistent with a pro-choice position, except for the caveat that if all rights are only antes not trumps, then the freedom to choose abortion is itself not exactly a right. It is subject to the same opportunities for alteration, even denial, as is any other private action that carries public impact. Those opportunities are defined by the vicissitudes and protections of democratic discourse. Ultimately, As Chambers (1993) argues, "rights are only as strong as the shared understand-

ings that underpin them" (p. 234). For a treatment of the abortion issue I take to be sympathetic to my position, see L. W. Sumner (1981), *Abortion and Moral Theory* (Princeton: Princeton University Press).

33. *Federalist # 10.*

34. This interpretation I take to be largely derivative from other civic republican interpretations of the US founding. Prominent among them are, of course, those of Garry Wills, Gordon Wood, and Bernard Bailyn. For an intriguing argument of how more fully to bring the civic republican interpretation in line with the liberal, pluralist rendering offered by Louis Hartz, among many others, see Michael Lienesch (1988), *New Order of the Ages: Time, the Constitution, and the Making of Modern American Political Thought* (Princeton: Princeton University Press).

35. Letter to William S. Smith (1787).

36. Glendon (1991), pp. 48–50.

37. Samuel D. Warren and Louis D. Brandeis (1890), "The Right to Privacy," *Harvard Law Review* 4:193. Quoted in Glendon (1991), p. 51.

38. *Coppage v. Kansas* 236 US 1 (1915). Quoted in Glendon (1991), p. 55.

39. One exception to this extension concerns homosexual couples. In *Bowers v. Hardwick*, 478 US 186 (1986), the Court refused to extend the right of privacy to homosexuals engaging in acts of sodomy, defined broadly to include oral or anal sex. See Susan Okin, "Humanist Liberalism," in Nancy L. Rosenblum (1989), *Liberalism and the Moral Life* (Cambridge: Harvard University Press), pp. 39–53, for a discussion of the expansion of the definition of the right to privacy within legal opinion.

40. Thomson (1986), p. 117.

41. Thomson (1986), pp. 126–27.

42. Waldron (1987), p. 184. The quotation of Marx is from "On the Jewish Question," reprinted in Waldron.

43. Held (1984), p. 216ff. In so interpreting the right of expression, Held clearly treats it as a "cluster right."

44. Held (1984), pp. 226–27, 216.

45. See my (1996) "Acts of Democracy: Reconceptualizing Politics, Participation, and Competence," in *The Public Perspective* 7:40–44, June/July.

46. Michael Ignatieff (1984), *The Needs of Strangers* (New York: Penguin), p. 13.

47. Glendon (1991), chs. 4, 5.

48. *New York Times*, March 14, 1995, "Deadly Nuclear Waste Piles Up With No Clear Solution at Hand," p. C1.

49. Alan Gewirth, "The Basis and Content of Human Rights," in Winston (1989), pp. 183–84.

50. Meyers (1985) acknowledges only four inalienable rights: the right to life, the right to personal liberty, the right to benign treatment, and the right to satisfaction of basic needs for survival (p. 52).

51. See my (1982) *Community Without Coercion: Getting Along in the Minimal State* (Newark: University of Delaware Press), ch. 5 for a thorough exploration of the literature on interest.

52. See Brian Barry (1965), *Political Argument* (New York: Humanities Press).

53. S. I. Benn (1960), "Interests in Politics," in *Meeting of the Aristotelian Society*, p. 135.

54. William Connolly (1974), *The Terms of Political Discourse* (Lexington, MA: D. C. Heath, p. 54.

55. Connolly (1974), p. 65.

56. Flathman (1976), pp. 188–89.

57. For a harrowingly accurate assessment of the new "clear and present danger" posed by the decommissioning of nuclear weapons in the absence of any plan for disposal of radioactive materials, see the *Chemical & Engineering News*, June 13, 1994, pp. 5–25.

5. *Risk and Responsibility*

1. Jeremy Waldron (1987), *Nonsense Upon Stilts* (London: Methuen), p. 195.

2. Larry May (1992), *Sharing Responsibility* (Chicago: University of Chicago Press), p. 34.

3. See Robert E. Goodin (1985), "Vulnerabilities and Responsibilities: An Ethical Defense of the Welfare State," *American Political Science Review* 79:775–86, especially pages 80ff.

4. See Held (1984), "Moral Theory and Moral Experience," in *Rights and Goods* (Chicago: University of Chicago Press).

5. For instance, in a recent article in *Ethics* (if not the most prominent journal of moral philosophy, then surely one of the top three), the authors recount numerous examples of dilemmas of moral responsibility, with the metaphorical titles of "assassin," "hero," "missile," "driver," and so forth. Although the piece is well argued, its point (concerning the relatedness of responsibility for consequences as for actions and omissions) could have been made more effectively, I think, had it been elucidated by instances of real events in which consequences, actions, and omissions raised subtle issues of responsibility for those involved. After all, the world is, for good or ill, full of flesh-and-blood instances of all four metaphors. See John Martin Fischer and Mark Ravizza (1991), "Responsibility and Inevitability," *Ethics* 101:258–78.

6. Held (1984), p. 43.

7. Ibid., p. 51.

8. P. F. Strawson (1974), "Freedom and Resentment," in Strawson, *Freedom and Resentment and Other Essays* (London: Methuen), pp. 1–25.

9. For a powerful discussion of the responsibility/criminal negligence issues in the case of the Bhopal disaster, see Ward Morehouse (1993), "The Ethics of Industrial Disasters in a Transnational World: The Elusive Quest for Justice and Accountability in Bhopal," *Alternatives* 18:475–504.

10. I will be relying primarily on his 1985 book, *Protecting the Vulnerable: A Reanalysis of Our Social Responsibilities* (Chicago: University of Chicago Press).

11. Marion Smiley (1992), *Moral Responsibility and the Boundaries of Community* (Chicago: University of Chicago Press). Dewey reference is from *Logic of Inquiry* (1938) (New York), p. 499, cited in Smiley, p. 25.

12. Goodin (1985) in his book's introduction presents an admirable list and discussion of sources throughout the history of moral philosophy that presumes, in his words, "a prejudice in favor of the people who stand in some special relationship to us." His list includes Sir David Ross, Bernard Williams, Derek Parfit, Joel Feinberg, John Rawls, and others. See p. 6ff.

13. I am aware that the federal building in Oklahoma City was not a true skyscraper because it had only 10 floors. Still, it was a building that could only be built in the twentieth century under modern construction techniques. Unlike hypothetical examples, real events are not always as conveniently descriptive of every element as one would like. If it helps, one could use as an example for this technology the bombing at the World Trade Center in New York a year earlier. There is no shortage—sad to say—of good examples.

14. For discussions of the different risk perceptions of catastrophic versus chronic or noncatastrophic risk see Baruch Fischhoff, et al. (1981), Acceptable Risks (Cambridge: Cam-

bridge University Press); K. S. Shrader-Frechette (1991), *Risk and Rationality* (Berkeley: University of California Press); Fischhoff, et al. (1978), "How Safe Is Safe Enough?" *Policy Sciences* 9:2; and E. Lawless (1977), *Technology and Social Shock* (New Brunswick, NJ: Rutgers University Press).

15. I am grateful (although somewhat warily so) to Anne L. Hiskes for this point.

16. Seattle and San Francisco both passed such citizen initiatives. For a discussion of them, see Robert E. McGinn (1994), "Technology, Demography, and the Anachronism of Traditional Rights," *Journal of Applied Philosophy* 11:57–70.

17. David Campbell (1994), "The Deterritorialization of Responsibility: Levinas, Derrida, and Ethics After the End of Philosophy," *Alternatives* 19:455–84.

18. Lynton K. Caldwell makes a case for what he calls "social responsibility" in an interesting piece that takes the notion of responsibility to the level of society itself as agent. This is, as far as I know, a different and new idea that deserves further exploration and elaboration. See Caldwell (1994), "Conditions for Social Responsibility," *American Behavioral Scientist* 38:172–92.

19. Campbell (1984), p. 463.

20. All three criteria are taken from Joel Feinberg (1970), *Doing and Deserving* (Princeton, NJ: Princeton University Press), pp. 30–33, quotation from p. 31.

21. Ibid., p. 33.

22. Thus, for instance, Michael Simon argues that tort law is an inappropriate vehicle for resolving responsibility and liability issues in many risk cases. Causation of harm is often difficult to prove for the reasons I have already discussed, and, furthermore, exposure to risk is difficult to define as actual harm. Thus, Simon argues, compensation decisions should not be made in civil court for the reasons already discussed in chapter four. Furthermore, Simon argues that risk exposure should not be a matter of private responsibility of persons or, in many cases, corporations. Instead, compensation should be a matter of public policy. Simon's argument converges with mine at many points, both here and in the next chapter where I discuss compensation for risk. See Michael Arthur Simon (1992), "Causation, Liability and Toxic Risk Exposure," *Journal of Applied Philosophy* 9:35–44.

23. Smiley (1992), p. 255.

24. See *The New York Times*, May 17, 1995, pp. A1, A14.

25. French's indictment of Lockean theory here is particularly persuasive. As he points out, contract theory has many problems with conceptualizing the obligations to future generations, but none more so than that involved in Locke's theory of property. Because Locke's theory does not even include a notion of waste, he is of little help in solving our problems arising from the disposal of hazardous waste without completely trampling on the interests of future generations. See Peter A. French (1992), *Responsibility Matters* (Lawrence University Press of Kansas), p. 97.

26. Ibid., p. 100.

27. Peter A. French (1984), *Collective and Corporate Responsibility* (New York: Columbia University Press), p. 32.

28. Ibid.

29. Ronald Dworkin also uses this approach in his example of corporate responsibility as a metaphor for community responsibility. What he calls the "deep personification" of corporate entities in law allows us to view them (and therefore the community as a whole) as moral entities, even though "the community has no independent metaphysical existence, that it itself is a creature of the practices of thought and language in which it figures." Dworkin (1986), *Law's Empire* (Cambridge, MA: Belknap), p. 171.

30. Larry May (1987), *The Morality of Groups* (South Bend, IN: University of Notre Dame Press), p. 84.

31. Dennis F. Thompson (1980), "Moral Responsibility of Public Officials: The Problem of Many Hands," *American Political Science Review* 74:905–16, p. 907.

32. Ibid., pp. 908–9.

33. May (1992), p. 38.

34. May (1987), p. 77.

35. This is important for Thompson in order to enhance the public accountability of identifiable elected and appointed officials. It is crucial for May in order to assign blame to the collectivity as a whole if it turns out that the leaders are especially responsible.

36. We will explore this second instance of responsibility a bit further on in this chapter.

37. The example is introduced by Peter Singer (1982), "Famine, Affluence, and Morality," *Philosophy and Public Affairs* 1:229–43. Singer makes a similar usage of the example to demonstrate the responsibility of the first world for Third World hunger.

38. My position differs here from that of many philosophers who are willing to ascribe collective responsibility but do not distribute it to individuals. For instance, McKinsey argues that in the instance of a disastrous flood, victims can hold the community responsible for failing to organize itself sufficiently to provide all the different services that need to be provided, but they cannot so hold individual persons responsible for any particular service. My use of shared responsibility (following May) allows the second ascription as well. See M. McKinsey (1981), "Obligations to the Starving," *Nous* 15:309–23. Cited in Goodin (1985), p. 137.

39. Virginia Held (1970), "Can a Random Collection of Individuals be Morally Responsible?" *Journal of Philosophy* 67:471–80, especially p. 479.

40. Annette C. Baier (1993), "How Can Individualists Share Responsibility?" *Political Theory* 21:228–48, p. 244.

41. Baier (1993), pp. 244, 245.

42. Goodin (1985), *Protecting*, p. 110. See also Goodin (1985), "Vulnerabilities."

43. Goodin (1985), *Protecting*, p. 11.

44. Ibid., p. 13, ch. three.

45. Smiley (1992), p. 255.

46. Ibid., pp. 10, 76.

47. Ibid., p. 55.

48. Ibid., pp. 174–77.

49. The issue of moral luck and its place in moral theory is an important one and has been commented on at least since Locke. For some contemporary renderings, see for instance Bernard Williams (1981), *Moral Luck* (Cambridge: Cambridge University Press); also Feinberg (1970), esp. ch. 2; Thomas Nagel, "Moral Luck," in his (1978) *Mortal Questions* (Oxford: Oxford University Press); and Arthur Ripstein (1994), "Equality, Luck, and Responsibility," *Philosophy and Public Affairs* 23:3–23.

50. Smiley (1992), p. 235.

51. See Simon (1992).

52. Ripstein (1994), p. 7.

6. *Democratic Politics and Participatory Risks*

1. Terence Ball and J. G. A. Pocock (1988), *Conceptual Change and the Constitution* (Lawrence: University Press of Kansas), p. 20.

2. Of course, political theorists must accept that they have a function in mediating such "mundane" matters as risk policy disputes and decision. Such acceptance might be difficult to achieve, given the rarefied realm that academic political theory has occupied for at least the past decade. For a discussion about the "strange silence" of political theory on important

issues of contemporary politics, see the symposium in *Political Theory* (1995) 23:635–88, beginning with the article by Jeffrey C. Isaac, "The Strange Silence of Political Theory."

3. This is, of course, the dilemma of individual liberty that Rousseau identified when acknowledging the need for the Legislator to "change human nature" before the society of the social contract could become possible. Social interests at some point do indeed presume a degree of self-denial or other-directedness that gives witness to the fact that self-identity has changed away from purely internal motivation and attached itself to the welfare of the larger community. The greatness of Rousseau's concept of the Legislator is the recognition that only a force external to the individual can interrupt the cycle of self-absorption and psychic dependence on others endemic to modern social life. Risks as an emergent force may provide the needed impetus.

4. Connolly (1994) notes the essential role of "boundaries" in the establishment of identity. He says, "Boundaries provide preconditions of identity, individual agency and collective action; but they also close off possibilities of being that might otherwise flourish." "Tocqueville, Territory and Violence," *Theory, Culture & Society* 11:19–41, p. 19. Because risks and their distribution often focus on the boundaries between individuals and their environment, the blurring of those boundaries makes for a particularly contestable form of identity politics, where what is at issue, is, as Connolly continues, the question of "how to come to terms with the ambiguity of boundaries. How to fight against their sacrifices and violences without sacrificing their advantages altogether."

5. William E. Connolly (1991), *Identity/Difference: Democratic Negotiations of Political Paradox* (Ithaca, NY: Cornell University Press), p. 64. Identity politics presents a growing body of literature spanning wide areas of political science and theory from feminism to international relations. Some examples include Seyla Benhabib (1987), " The Generalized and Concrete Other," in S. Benhabib and Drucilla Cornell, eds. (1987), *Feminism as Critique* (Minneapolis: University of Minnesota Press); also, Benhabib (1992), *Situating the Self: Gender, Community, and Postmodernism in Contemporary Ethics* (New York: Routledge); Anna Elisabetta Galeotti (1993), "Citizenship and Equality: The Place for Toleration," *Political Theory* 21:4; Kenneth R. Hoover (1997), *The Power of Identity* (Chatham, NJ: Chatham House Publishers, Inc.); Will Kymlicka (1989), *Liberalism, Community, and Culture* (Oxford: Clarendon); Norma Claire Moruzzi (1994), "A Problem with Headscarves: Contemporary Complexities of Political and Social Identity," *Political Theory* 22:4; Anne Norton (1988), *Reflections on Political Identity* (Baltimore: Johns Hopkins University Press); John Rawls (1993), *Political Liberalism* (New York: Columbia University Press); Charles Taylor (1989), *Sources of the Self: The Making of Modern Identity* (Cambridge: Harvard University Press); Iris Young (1990), *Justice and the Politics of Difference* (Princeton, NJ: Princeton University Press). For an excellent review and critique of several of these and other sources, see Mark Reinhardt (1995), "Look Who's Talking: Political Subjects, Political Objects, and Political Discourse in Contemporary Theory," *Political Theory* 23:689–719.

6. Taylor (1989), p. 28.

7. Galeotti (1993). See also a response to Galeotti, with rejoinder, by Moruzzi (1994).

8. Galeotti (1993), p. 586.

9. For a classic discussion and critique of liberalism's reliance on the construct of an abstract individual see Marx's *On the Jewish Question*. More contemporary works include Seyla Benhabib (1986), "The Generalized and Concrete Other," in Seyla Benhabib and Drucilla Cornell, eds., *Feminism as Critique* (Minneapolis: University of Minnesota Press), pp. 77–95; Richard P. Hiskes (1982), *Community Without Coercion* (Newark: University of Delaware Press); Steven Lukes (1973), *Individualism* (New York: Harper and Row); Robert Paul Wolff (1968), *The Poverty of Liberalism* (Boston: Beacon); and Iris Marion Young (1986), "Impartiality and the Civic Public," in Benhabib and Cornell, pp. 56–76.

10. See Robert E. Goodin (1982), *Political Theory and Public Policy* (Chicago: University of Chicago Press), especially chs. 8, 10; also, Francis O. Adeola (1994), "Environmental Hazards, Health, and Racial Inequity in Hazardous Waste Distribution," *Environment and Behaviour* 26:99–126; John S. Dryzek (1987), *Rational Ecology: Environment and Political Economy* (Oxford: Blackwell); Robyn Eckersley (1992), *Environmentalism and Political Theory: Toward an Ecocentric Approach* (Albany: State University of New York Press); Richard Hofrichter (1993), *Toxic Struggles: The Theory and Practice of Environmental Justice* (Philadelphia: New Society Publishers); William Ophuls (1977), *Ecology and the Politics of Scarcity* (San Francisco: W. H. Freeman).

11. See Enrique Gonzalez-Manet (1988), *The Hidden War of Information*, Laurien Alexandre, trans. (Norwood, NJ: Ablex Publishing); Elihu Katz and George Wedell (1977), *Broadcasting in the Third World* (Cambridge: Harvard University Press); Kaarle Nordenstreng and Herbert I. Schiller, eds. (1979), *National Sovereignty and International Communication* (Norwood, NJ: Ablex Publishing); Michael Robinson (1983), *Over the Wire and on TV* (New York: Russell Sage); and Herbert Schiller (1976), *Communication and Cultural Domination* (White Plains, NY: International Arts and Science Press). For a dissenting view see Ithiel de Sola Pool (1990), *Technologies Without Boundaries* (Cambridge: Harvard University Press), but also my review in *American Political Science Review* 86:3, Sept. 1992.

12. Michael Ignatieff (1984), *The Needs of Strangers* (New York: Penguin), p. 139.

13. Connolly (1994), p. 32.

14. James Bohman (1995), "Public Reason and Cultural Pluralism: Political Liberalism and the Problem of Moral Conflict," *Political Theory* 23:2, p. 254.

15. Rawls (1993), p. xvi.

16. John Gray (1995), "Agonistic Liberalism," *Social Philosophy and Policy* 10:111–35.

17. Connolly (1991), pp. 65, 66.

18. Michael Walzer (1989), "Citizenship," in Terence Ball, James Farr, and Russell L. Hanson (1989), *Political Innovation and Conceptual Change* (Cambridge: Cambridge University Press); p. 215.

19. Ibid., p. 218. J. Donald Moon further emphasizes the liberal emphasis on the private realm as the locale of normative concerns. He notes the "liberal problematic." "Liberal political theory can be interpreted as an effort to answer the question How can people who disagree on important and enduring questions regarding the ends and purposes of life come to live together under rules that they can all accept as just and reasonable . . . ? The basic liberal response [is] to erect a separation between the public sphere, in which matters are subject to collective, authoritative decision and control, and the private sphere, in which individuals are free to control and direct their own behavior." "Theory, Citizenship, and Democracy," in George E. Marcus and Russell L. Hanson (1993), *Reconsidering the Democratic Public* (University Park: Pennsylvania State University Press), pp. 211–22, p. 219.

20. Pamela Johnston Conover, Stephen T. Leonard, and Donald D. Searing (1993), "Duty is a Four-Letter Word: Democratic Citizenship in the Liberal Polity," in Marcus and Hanson, pp. 147–72.

21. Michael Walzer (1990), "The Communitarian Critique of Liberalism," *Political Theory* 18:6–23, quoted in Mary G. Dietz (1993), "In Search of a Citizen Ethic," in Marcus and Hanson, p.176. Dietz argues, contra Conover et al., that it is the denial of the notion of common good among liberal citizens, not their rejection of duties, that separates liberals from their communitarian critics.

22. Philip J. Frankenfeld (1992), "Technological Citizenship: A Normative Framework for Risk Studies," *Science, Technology, & Human Values* 17:459–484, p. 459. For an exegesis and discussion of Frankenfeld's conception of citizenship, see Andrew D. Zimmerman

(1995), "Toward a More Democratic Ethic of Technological Governance," *Science, Technology, and Human Values* 20:86–107.

23. Frankenfeld, p. 459.

24. Ibid., p. 473. Frankenfeld argues that the use of knowledge in "TC" will lead to civic virtue in the sense of individuals realizing a growing responsibility to a "technological system" that itself grows to include "ever more remote persons affected by ever more abstract impacts of one's actions"(p. 475). Thus, he claims that TC encourages moral development in involving individuals in increasingly abstract rationalizations of morality, as described in Laurence Kohlberg's moral psychological theories. Of course, Kohlberg's whole theory is often criticized for relying overly much on a liberal view of the individual as "abstract." Thus, to the extent that what is needed is a nonliberal view of citizenship, Kohlberg's mark of approval is hardly reassuring. See Carol Gilligan (1982), *In a Different Voice: Psychological Theory and Women's Development* (Cambridge, MA: Harvard University Press).

25. Jean Bethke Elshtain (1995), *Democracy on Trial* (New York: Basic), p. 30.

26. For an interesting example of assigning responsibility to groups for the fact of discrimination and sexual harassment, see Larry May (1992), *Sharing Responsibility* (Chicago: University of Chicago Press).

27. Mary G. Dietz (1993), "In Search of a Citizen Ethic," in Marcus and Hanson, pp. 173–85, p. 180, emphasis Dietz's.

28. Connolly (1991), pp. 198–99.

29. Mary Gibson (1985), "Consent and Autonomy," in Mary Gibson, ed., *To Breathe Freely: Risk, Consent, and Air* (Totowa, NJ: Rowman & Allanheld), p. 145. The Piaget work cited by Gibson is (1965), *The Moral Judgment of the Child* (New York: The Free Press).

30. Gibson, p. 151. For an application of de Tocqueville's approach to civic participation as the true hope for democratic citizenship, and for democracy's future, see Elshtain (1995) and Michael Sandel (1996), *Democracy's Discontent* (Cambridge, MA: Belknap). See also my discussion of both their books in the context of a general appraisal of democracy's current prospects in light of technological developments in R. P. Hiskes (1996), "Acts of Democracy: Reconceptualizing Politics, Participation, and Competence," in *Public Perspective* 7:40–44.

31. Carol C. Gould (1988), *Rethinking Democracy; Freedom and Social Cooperation in Politics, Economy, and Society* (Cambridge: Cambridge University Press), p. 19.

32. Ibid., p. 105.

33. Gould herself agrees that technology offers one of the most severe challenges to the positive freedom characteristic of democracy. Although only an instrument, not itself an autonomous force, modern technology nevertheless, according to Gould, "makes possible the intensification of control by few over the work activity of large numbers who require this technology as a means of their labor." Thus, "[t]echnology used in these ways, violates not only the value of freedom, but also the values of equality and reciprocity." Ibid., pp. 266–67, 268.

34. Ibid., p. 132.

35. Benjamin Barber (1984), *Strong Democracy* (Berkeley: University of California Press); also, Benjamin Barber (1988), *The Conquest of Politics* (Princeton, NJ: Princeton University Press); John S. Dryzek (1990), *Discursive Democracy* (Cambridge: Cambridge University Press); James S. Fishkin (1991), *Democracy and Deliberation* (New Haven, CT: Yale University Press); Amy Gutmann (1993), "The Disharmony of Democracy," in John W. Chapman and Ian Shapiro, eds., NOMOS XXXV: *Democratic Community* (New York: New York University Press); Connolly (1991); and Thomas L. Pangle (1992), *The Ennobling of Democracy* (Baltimore: Johns Hopkins University Press). Connolly also uses the term "territorial democracy" to characterize the conflict over identity characteristic of his view of agonistic politics.

36. Dryzek (1990), p. 6.

37. See the discussion of civic virtue and the rule of law in Pangle (1992), p.107ff. See also Paul Stern (1997), "The Rule of Wisdom and the Rule of Law in Plato's *Statesman*," *American Political Science Review* 91:264–76, June, for an extended discussion of the concept of *phronesis* in Greek thought.

38. Pangle, p. 120.

39. See Helen Longino (1986), "Science Overrun: Threats to Freedom from External Control," in Malcolm L. Goggin, ed., *Governing Science and Technology in a Democracy* (Knoxville: University of Tennessee Press), pp. 57–74.

40. Barber (1988), pp. 200, 209, 210.

41. Barber (1988), p. 200.

42. Barber (1988), p. 201.

43. Dryzek (1990), pp. 4–6.

44. Dryzek (1990), p. 12.

45. Dryzek (1990), p. 15, 20.

46. Connolly (1991), p. 200.

47. Dryzek (1990), p. 41.

48. Increasingly, political scientists are realizing that participation—or lack of it—is not only a function of interest and political interests. Indeed, in a recent study, three prominent scholars of political behavior conclude that "the resources of time, money, and (civic) skills" are as powerful predictors of political participation in America as economic interest. See Henry E. Brady, Sidney Verba, and Kay Lehman Schlozman (1995), "Beyond SES: A Resource Model of Political Participation," *American Political Science Review* 89:271–94.

49. As Carole Pateman (1979) states, "consent has to be seen as something that arises from the acts of citizens." *The Problem of Political Obligation* (Chichester: John Wiley & Sons), p. 88.

50. Gutmann (1993), p. 134.

51. I am indebted to Professor Any Gutmann for this point.

52. Barber (1988), p. 206.

53. J. Donald Moon (1993), "Theory, Citizenship, and Democracy," in Marcus and Hanson, p. 213. Note also Jennifer L. Hochschild's cautiously optimistic reconstruction of voter inconsistency in "Disjunction and Ambivalence in Citizens' Political Outlooks," also in Marcus and Hanson.

54. Moon, p. 214.

55. For a definition of hedgehog and fox-like thinking, see Isaiah Berlin (1979), *Russian Thinkers* (New York: Penguin). Moon does a good job of summarizing the distinctions between the two cognitive styles.

56. Gutmann (1993), p. 143.

57. See Frank Fischer (1993), "Citizen Participation and the Democratization of Policy Expertise: From Theoretical Inquiry to Practical Cases," *Policy Sciences* 26:165–87; John Forester (1993), *Critical Theory, Public Policy and Planning Practice: Toward a Critical Pragmatism* (Albany: State University of New York Press); Ray Kemp (1985), "Planning, Public Hearings, and the Politics of Discourse," in John Forester, ed., *Critical Theory and Public Life* (Cambridge, MA: MIT Press); Ronald J. Fisher (1983), "Third Party Consultation as a Method of Intergroup Conflict Resolution," *Journal of Conflict Resolution* 27:301–34; Roger Fisher and William Ury (1981), *Getting to Yes* (Boston: Houghton Mifflin); Sam Gusman (1981), "Policy Dialogue," *Environmental Comment* November 14–16; Phillip J. Harter (1982), "Negotiating Regulations: A Cure for Malaise," *Georgetown Law Journal* 71:1–118; James A. Wall, Jr. (1981), "Mediation: An Analysis, Review, and Proposed Research," *Journal of Conflict Resolution* 25:157–80. For a general defense of referendums in an otherwise in-

ternational politics context, see John T. Rourke, Richard P. Hiskes, and Cyrus Ernesto Zirakzadeh (1992), *Direct Democracy and International Politics* (Boulder, CO: Lynne Rienner Publishers).

58. See Bruce Jennings (1986), "Representation and Participation in the Democratic Governance of Science and Technology," for a discussion of the difference between procedural and substantive democracy when it comes to policy making for science and technology, in Goggin (1986).

59. Gutmann (1993), p. 128.

Bibliography

Ackerman, William. *Social Justice in the Liberal State.* New Haven: Yale University Press, 1980.

Adeola, Francis O. "Environmental Hazards, Health, and Racial Inequity in Hazardous Waste Distribution." *Environment and Behaviour* 26:99–126, 1994.

Aiken, Henry D., ed. *Hume's Moral and Political Philosophy.* New York: Hafner, 1972.

Appelbaum, Paul S., Charles W. Lidz, and Alan Meisel. *Informed Consent: Legal Theory and Clinical Practice.* New York: Oxford, 1987.

Arendt, Hannah. *Between Past and Future.* New York: Viking, 1961.

Ashcraft, Richard. *Revolutionary Politics & Locke's Two Treatises of Government.* Princeton, NJ: Princeton University Press, 1986.

Bachrach, Peter. *The Theory of Democratic Elitism.* Washington, DC: University Press of America, 1980.

Bachrach, Peter, and Morton S. Baratz. "Two Faces of Power." *American Political Science Review* 56:947–52, 1962.

Baier, Annette C. "How Can Individualists Share Responsibility?" *Political Theory* 21:228–48, 1993.

Ball, Terence, James Farr, and Russell L. Hanson. *Political Innovation and Conceptual Change.* Cambridge: Cambridge University Press, 1989.

Ball, Terence, and J. G. A. Pocock. *Conceptual Change and the Constitution.* University Press of Kansas, 1988.

Barber, Benjamin. *The Conquest of Politics.* Princeton, NJ: University Press, 1988.

———. *Strong Democracy.* Berkeley: University of California Press, 1984.

Barry, Brian. *Political Argument.* New York: Humanities Press, 1965.

———, ed. *Power and Political Theory.* London: John Wiley and Sons, 1977.

———. "Warrender and His Critics." *Philosophy* 43:117–37, 1968.

Bayles, M. *Professional Ethics.* Belmont, CA: Wadsworth, 1981.

Beauchamp, T. L., and J. F. Childress. *Principles of Biomedical Ethics.* New York: Oxford, 1989.

Beck, Ulrich. *Risk Society.* London: Sage, 1992.

Bedau, Hugo. "The Right to Life." *Monist* 52:560–72.

Bellah, Robert, et al. *Habits of the Heart.* Berkeley: University of California Press, 1985.

Benhabib, Seyla. *Situating the Self.* New York: Routledge, 1992.

Benhabib, Seyla, and Drucilla Cornell, eds. *Feminism as Critique*. Minneapolis: University of Minnesota Press, 1987.

Benn, S. I. "Interests in Politics." *Meeting of the Aristotelian Society*. 1960.

Berlin, Isaiah. *Four Essays on Liberty*. Oxford: Oxford University Press, 1969.

———. *Russian Thinkers*. New York: Penguin, 1979.

Blank, Robert. *The Political Implications of Human Genetic Technology*. Boulder, CO: Westview, 1981.

Bohman, James. "Public Reason and Cultural Pluralism: Political Liberalism and the Problem of Moral Conflict." *Political Theory* 23:253–79, 1995.

Brady, Henry E., Sidney Verba, and Kay Lehman Schlozman. "Beyond SES: A Resource Model of Political Participation." *American Political Science Review* 89:271–94, 1995.

Brodbeck, May, ed. *Readings in the Philosophy of the Social Sciences*. New York: Macmillan, 1968.

Bronowski, Jacob. *The Ascent of Man*. Boston: Little, Brown, 1973.

Bryant, B., and P. Mohai, eds. *Race and the Incidence of Environmental Hazards: A Time for Discourse*. Boulder, CO: Westview, 1992.

Bullard, R. D. *Dumping in Dixie: Race, Class, and Environmental Decisionmaking*. Boulder, CO: Westview, 1994.

Burger, Edward J., Jr., ed. *Risk*. Ann Arbor: University of Michigan Press, 1990.

Caldwell, Lynton K. "Conditions for Social Responsibility." *American Behavioral Scientist* 38:172–92, 1994.

Campbell, David. "The Deterritorialization of Responsibility: Levinas, Derrida, and Ethics After the End of Philosophy." *Alternatives* 19:455–84, 1994.

Carr, Craig L. "Coercion and Freedom." *American Philosophical Quarterly* 25:59–67, 1988.

———. "Tacit Consent." *Public Affairs Quarterly* 4:335–45, 1990.

Chambers, Simone. "Talking About Rights: Discourse Ethics and the Protection of Rights." *Journal of Philosophy* 1:229–49, 1993.

Chapman, John W., and Ian Shapiro. *NOMOS XXXV: Democratic Community*. New York: NYU PRess, 1993.

Cole, G. D. H. *Rousseau: The Social Contract and Discourses*. New York: Everyman, 1950.

Cole, Sam. "Cultural Technological Futures." *Alternatives* 15:377–400, 1990.

Connolly, William E. *Identity/Difference: Democratic Negotiations of Political Paradox*. Ithaca, NY: Cornell University Press, 1991.

———. *The Terms of Political Discourse*. Lexington, MA: D. C. Heath, 1974.

———. "Tocqueville, Territory, and Violence." *Theory, Culture, & Society* 11:19–41, 1994.

Conrad, J., ed. *Society, Technology, and Risk*. New York: Academic Press, 1980.

Curry, M., and Larry May. *Professional Responsibility for Harmful Actions*. Dubuque, IA: Kendal/Hunt, 1984.

Davis, Charles E. "Public Involvement in Hazardous Waste Siting Decisions." *Polity* 19:296–304, 1986.

Dawkins, Richard. *The Selfish Gene*. New York: Oxford University Press, 1976.

Derrida, Jacques. *Acts of Literature*. New York: Routledge, 1992.

Dewey, John. *The Quest for Certainty: A Study of the Relation and Knowledge and Action*. New York: Minton, Balch, 1929.

Dietz, Thomas M., and Robert W. Rycroft. *The Risk Professionals*. New York: Russell Sage, 1987.

Douglas, Mary, and Aaron Wildavsky. *Risk and Culture*. Berkeley: University of California Press, 1982.

Dryzek, John. *Discursive Democracy*. Cambridge: Cambridge University Press, 1990.

———. *Rational Ecology: Environment and Political Economy*. Oxford: Blackwell, 1987.
Dworkin, Gerald, ed. *The Theory and Practice of Autonomy*. Cambridge: Cambridge University Press, 1988.
Dworkin, Ronald. *Law's Empire*. Cambridge, MA: Belknap, 1986.
———. *Taking Rights Seriously*. Cambridge, MA: 1977.
Eckersley, Robyn. *Environmentalism and Political Theory*. Albany: SUNY Press, 1992.
Elshtain, Jean Bethke. *Democracy on Trial*. New York: Basic, 1995.
Faden, Ruth, and Thomas Beauchamp. *A History and Theory of Informed Consent*. New York: Oxford, 1986.
Feinberg, Joel. *Doing and Deserving*. Princeton, NJ: Princeton University Press, 1970.
———. *Rights, Justice, and the Bounds of Liberty*. Princeton, NJ: Princeton University Press, 1980.
Fischer, Frank. "Citizen Participation and the Democratization of Policy Expertise." *Policy Sciences* 26:165–87, 1993.
Fischer, John Martin, and Mark Ravizza. "Responsibility and Inevitability." *Ethics* 101:258–78, 1991.
Fischhoff, Baruch, Sarah Lichtenstein, Paul Slovic, Steven L. Derby, Ralph L. Keeney. *Acceptable Risk*. Cambridge: Cambridge University Press, 1981.
Fischhoff, Baruch, Paul Slovic, Sarah Lichtenstein, Steven Read, Barbara Combs. "How Safe Is Safe Enough?" *Policy Sciences* 9:127–52, 1978.
Fisher, Roger, and William Ury. *Getting to Yes*. Boston: Houghton Mifflin, 1981.
Fisher, Ronald J. "Third Party Consultation as a Method of Intergroup Conflict Resolution." *Journal of Conflict Resolution* 27:301–34, 1983.
Fishkin, James S. *Democracy and Deliberation*. New Haven, CT: Yale University Press, 1991.
Flathman, Richard E. *Concepts in Social and Political Philosophy*. New York: Macmillan, 1973.
———. *The Practice of Political Authority*. Chicago: University of Chicago Press, 1980.
———. *The Practice of Rights*. Cambridge: Cambridge University Press, 1976.
Flavin, Catherine, and Regina Dougherty. "Science And Citizenship at the NIC." *Public Perspective* 7:46–9, 1996.
Foege, William H. "Plagues: Perceptions of Risk and Social Responses." *Social Research* 55:331–42, 1988.
Forester, John. *Critical Theory and Public Life*. Cambridge, MA: MIT Press, 1985.
———. *Critical Theory, Public Policy and Planning Practice*. Albany: SUNY Press, 1993.
Foucault, Michel. *Discipline and Punish*. New York: Pantheon, 1977.
———. *The History of Sexuality*. New York: Pantheon, 1976.
Frankenfeld, Philip J. "Technological Citizenship: A Normative Framework for Risk Studies." *Science, Technology, & Human Values* 17:459–84, 1992.
French, Peter A. *Collective and Corporate Responsibility*. New York: Columbia University Press, 1984.
———. *Responsibility Matters*. Lawrence: University Press of Kansas, 1992.
Friedrich, Carl, ed. *NOMOS I: Authority*. Cambridge, MA: Harvard University Press, 1958.
Galeotti, Anna Elisabetta. "Citizenship and Equality: The Place for Toleration." *Political Theory* 21:585–605, 1993.
Galston, William A. "Defending Liberalism." *American Political Science Review* 76:621–29, 1982.
———. *Liberal Purposes*. Cambridge: Cambridge University Press, 1991.
Gibson, Mary. *To Breathe Freely: Risk, Consent, and Air*. Totowa, NJ: Rowman & Allanheld, 1985.
Gilbert, Margaret. "Agreements, Coercion, and Obligation." *Ethics* 103:679–706, 1993.

Gilligan, Carol. *In a Different Voice: Psychological Theory and Women's Development.* Cambridge, MA: Harvard University Press, 1982.

Gilmour, Robert S., and Alexis A. Halley. *Who Makes Public Policy: The Struggle for Control Between Congress and the Executive.* Chatham, NJ: Chatham House, 1994.

Glendon, Mary Ann. *Rights Talk.* New York: Free Press, 1991.

Goggin, Malcolm L., ed. *Governing Science and Technology in a Democracy.* Knoxville: University of Tennessee Press, 1986.

Gonzalez-Manet, Enrique. *The Hidden War of Information.* Norwood, NJ: Ablex Publishing, 1988.

Goodin, Robert E. *Political Theory and Public Policy.* Chicago: University of Chicago Press, 1982.

———. *Protecting the Vulnerable.* Chicago: University of Chicago Press, 1985.

———. "Vulnerabilities and Responsibilities: An Ethical Defense of the Welfare State." *American Political Science Review* 79:775–86, 1985.

Gorovitz, Samuel, ed. *Mill: Utilitarianism.* Indianapolis: Bobbs-Merrill, 1971.

Gould, Carol C. *Rethinking Democracy: Freedom and Social Cooperation in Politics, Economy, and Society.* Cambridge: Cambridge University Press, 1988.

Gray, John. "Agonistic Liberalism." *Social Philosophy and Policy* 10:111–35, 1995.

Gunderson, Martin. "Justifying a Principle of Informed Consent: A Case Study in Autonomy-Based Ethics." *Public Affairs Quarterly* 4:249–65, 1990.

Gusman, Sam. "Policy Dialogue." *Environmental Comment* Nov.: 14–16, 1981.

Gutmann, Amy, and Dennis Thompson. *Democracy and Disagreement.* Cambridge, MA: Belknap, 1996.

Habermas, Jurgen. *The Theory of Communicative Action.* Boston: Beacon, 1984.

Hanser, Matthew. "Harming Future People." *Philosophy & Public Affairs* 19:47–70, 1990.

Harris, R. Baine. *Authority: A Philosophical Analysis.* Tuscaloosa: University of Alabama Press, 1976.

Harter, Phillip J. "Negotiating Regulations: A Cure for Malaise." *Georgetown Law Journal* 71:1–118, 1982.

Held, Virginia. "Can a Random Collection of Individuals be Morally Responsible?" *Journal of Philosophy* 67:471–80, 1970.

———. *Rights and Goods.* Chicago: University of Chicago Press, 1984.

Higgins, Robert R. "Race & Environmental Equity: An Overview of the Environmental Justice Issue in the Policy Process. *Polity* 26:281–300, 1993.

Hileman, Bette. "Views Differ Sharply Over Benefits, Risks of Agricultural Biotechnology." *Chemical & Engineering News* August 21, 1995.

Hiskes, Anne L., and Richard P. Hiskes. *Science, Technology, and Policy Decisions.* Boulder, CO: Westview, 1986.

Hiskes, Richard P. "Acts of Democracy: Reconceptualizing Politics, Participation, and Competence." *Public Perspective* 7:40–4, 1996.

———. *Community Without Coercion: Getting Along in the Minimal State.* Newark: University of Delaware Press, 1980.

———. The Democracy of Risk." *Industrial Crisis Quarterly* 6:259–78, 1992.

Hofrichter, Richard. *Toxic Struggles: The Theory and Practice of Environmental Justice.* Philadelphia: New Society Publishers, 1993.

Hohfeld, Wesley Newcomb. *Fundamental Legal Conceptions.* New Haven, CT: Yale Univesity Press, 1964.

Hoover, Kenneth R. *The Power of Identity.* Chatham, NJ: Chatham House, 1997.

Ignatieff, Michael. *The Needs of Strangers.* New York: Penguin, 1984.

Isaac, Alan C. *Scope and Methods of Political Science.* Homewood, IL: Dorsey Press, 1969.

Isaacs, Jeffrey C. "The Strange Silence of Political Theory." *Political Theory* 23:636–52, 1995.

Jacob, Gerald. *Site Unseen: The Politics of Siting a Nuclear Waste Repository.* Pittsburgh: University of Pittsburgh Press, 1990.

Jasanoff, Sheila. "Bridging the Two Cultures of Risk Analysis." *Risk Analysis* 13:122–26, 1993.

Jones, Kathleen B. *Compassionate Authority: Democracy and the Representation of Women.* New York: Routledge, 1993.

Kasperson, Roger E. "Six Propositions on Public Participation and Their Relevance to Risk Communication. *Risk Analysis* 6:275–81, 1986.

———. "The Social Amplification of Risk: Progress in Developing an Integrative Framework." In Sheldon Krimsky and Dominic Golding, eds., *Social Theories of Risk.* Westport, CT: Praeger, 1992.

Kateb, George. *The Inner Ocean.* Ithaca, NY: Cornell University Press, 1992.

Katz, Elihu, and George Wedell. *Broadcasting in the Third World.* Cambridge: Harvard University Press, 1977.

Kraft, Michael E., and Bruce B. Clary, "Citizen Participation and the NIMBY Syndrome: Public Response to Radioactive Waste Disposal." *Western Political Quarterly* 44:299–328, 1991.

Kraft, Michael E., and Denise Scherberle. "Environmental Justice and the Allocation of Risk: The Case of Lead and Public Health." *Policy Studies Journal* 23:113–22, 1995.

Krimsky, Sheldon. "The Role of Theory in Risk Studies." In Sheldon Krimsky and Dominic Golding, eds. *Social Theories of Risk.* Westport, CT: Praeger, 1992.

Krimsky, Sheldon, and Dominic Golding, eds. *Social Theories of Risk.* Westport, CT. Praeger, 1992.

Kuhn, Thomas. *The Structure of Scientific Revolutions.* 2nd ed. Chicago: University of Chicago Press, 1970.

Kymlicka, Will. *Contemporary Political Philosophy: An Introduction.* Oxford: Clarendon Press, 1990.

———. *Liberalism, Community, and Culture.* Oxford: Clarendon Press, 1989.

Laffont, Jean-Jacques. "Regulation, Moral Hazard and Insurance of Environmental Risks." *Journal of Public Economics* 58:319–26, 1995.

Lash, Scott. "Reflexive Modernization: The Aesthetic Dimension." *Theory, Culture & Society* 10:1–23, 1993.

Lawless, E. *Technology and Social Shock.* New Brunswick, NJ: Rutgers University Press, 1977.

Leiss, William. "Assessing and Managing Risks." Review. *Policy Sciences* 25:341–49, 1992.

Leiss, William, and Christina Chociolko. *Risk and Responsibility.* Montreal: McGill-Queens University Press, 1994.

Lichtenstein, S., et al. "Judged Frequency of Lethal Events." *Journal of Experimental Psychology: Human Learning and Memory* 4:551–78, 1978.

Lienesch, Michael. *New Order of the Ages: Time, the Constitution, and the Making of Modern American Political Thought.* Princeton,NJ: Princeton University Press, 1988.

Linder, Stephen H. "Contending Discourses in the Electric and Magnetic Fields Controversy: The Social Construction of EMF Risk as a Public Problem." *Policy Sciences* 28:209–30, 1995.

Locke, John. *The Second Treatise of Government.* Thomas P. Peardon, ed. Indianapolis: Bobbs-Merrill, 1952.

Lowi, Theodore J. "Risk and Rights in the History of American Governments." In Peter J. Burger, Jr., ed., *Risk.* Ann Arbor: University of Michigan Press, 1990.

Luhmann, Niklas. *Risk: A Sociological Theory.* New York: Aldine DeGruyter, 1993.

Lukes, Steven. *Individualism.* New York: Harper and Row, 1973.

Machiavelli, Niccolò. *The Prince*. New York: New American Library, 1952.
MacIntyre, Alasdair. *After Virtue*. Notre Dame, IN: Notre Dame University Press, 1981.
———. *A Short History of Ethics*. New York: Macmillan, 1966.
MacLean, Douglas, ed. *Values at Risk*. Lanham, MD: Rowman & Littlefield, 1986.
MacPherson, C. B. *The Political Theory of Possessive Individualism*. Oxford: Oxford University Press, 1962.
———. *The Real World of Democracy*. Oxford: Oxford University Press, 1966.
Marcus, George E., and Russell L. Hanson. *Reconsidering the Democratic Public*. University Park: Pennsylvania State University Press, 1993.
Margolis, Howard. *Dealing With Risk*. Chicago: University of Chicago Press, 1996.
Matheny, Albert R. and Bruce A. Williams. "Knowledge v. NIMBY: Assessing Florida's Strategy for Siting Hazardous Waste Disposal Facilities. *Policy Studies Journal* 14:70–80, 1985.
May, Larry. *The Morality of Groups*. South Bend, IN: University of Notre Dame Press, 1987.
———. *Sharing Responsibility*. Chicago: University of Chicago Press, 1992.
McGinn, Robert E. "Technology, Demography, and the Anachronism of Traditional Rights." *Journal of Applied Philosophy* 11:57–70, 1994.
McKinsey, M. "Obligations to the Starving." *Nous* 15:309–23, 1981.
Meyers, Diana T. *Inalienable Rights: A Defense*. New York: Columbia University Press, 1985.
Mitchel, T. R., and W. G. Scott. "Leadership Failures, the Distrusting Public, and the Prospect of the Administrative State." *Public Administration Review* 47:445–52, 1987.
Morehouse, Ward. "The Ethics of Industrial Disasters in a Transnational World: The Elusive Quest for Justice and Accountability in Bhopal. *Alternatives* 18:475–504.
Morell, David, and Christopher Magorian. *Siting Hazardous Waste Facilities: Local Opposition and the Myth of Preemption*. Cambridge, MA: Balinger, 1982.
Morgan, M. Granger, Baruch Fischhoff, and Ann Bostrom. "Communicating Risk to the Public." *Environmental Science and Technology* 26:2048–56.
Moruzzi, Norma Claire. "A Problem with Headscarves: Contemporary Complexities of Political and Social Identity." *Political Theory* 22:653–72, 1994.
Nagel, Thomas. *Mortal Questions*. Oxford: Oxford University Press, 1978.
Nelkin, Dorothy. "Blunders in the Business of Risk." *Nature* 298:775–76, 1982.
———. *Controversy: Politics of Technical Decisions*. Beverly Hills, CA: Sage, 1979.
Nordenstreng, Kaarle, and Herbert I. Schiller, eds.*National Sovereignty and International Communication*. Norwood, NJ: Ablex Publishing, 1979.
Norton, Anne. *Reflections on Political Identity*. Baltimore: Johns Hopkins University Press, 1988.
Ophuls, William. *Ecology and the Politics of Scarcity*. San Francisco: W. H. Freeman, 1977.
Oppenheim, Felix E. *Political Concepts: A Reconstruction*. Chicago: University of Chicago Press, 1981.
Palmlund, Ingmar. "Social Drama and Risk Evaluation." In Sheldon Krimsky and Dominic Golding, eds., *Social Theories of Risk*. Westport, CT: Praeger, 1992.
Pangle, Thomas L. *The Ennobling of Democracy*. Baltimore: Johns Hopkins University Press, 1992.
Pateman, Carole. *Participation and Democratic Theory*. Cambridge: Cambridge University Press, 1970.
———. *The Problem of Political Obligation*. New York: John Wiley & Sons, 1979.
Paterson, Christopher J., and Richard N. L. Andrews. "Procedural and Substantive Fairness Risk Decisions: Comparative Risk Assessment Procedures." *Policy Studies Journal* 23: 85–95, 1995.

Pennock, J. Roland, and John W. Chapman. *NOMOS XXIX: Authority Revisited.* New York: NYU Press, 1987.

Petersen, James C., ed. *Citizen Participation in Science Policy.* Amherst: University of Massachusetts Press, 1984.

Pijawka, David K., and Alvin H. Mushkatel. "Public Opposition to the Siting of the High-Level Nuclear Waste Repository: The Importance of Trust." *Policy Studies Review* 10:180–94, 1991.

Pool, Ithiel de Sola. *Technologies Without Boundaries.* Cambridge: Harvard University Press, 1990.

Popper, Karl. *The Logic of Scientific Discovery.* New York: Harper & Row, 1934.

Price, Don K. *America's Unwritten Constitution: Science, Religion, and Political Responsibility.* Cambridge, MA: Harvard University Press, 1985.

Rabinow, Paul, ed. *The Foucault Reader.* New York: Pantheon, 1984.

Rawls, John. *Political Liberalism.* New York: Columbia University Press, 1993.

———. *A Theory of Justice.* Cambridge, MA: Harvard University Press, 1971.

Rayner, Steve. "Cultural Theory and Risk Analysis." In Sheldon Krimsky and Dominic Golding, eds., *Social Theories of Risk.* Westport, CT: Praeger, 1992.

———. *Risk in Cultural Perspective: Acting Under Uncertainty.* Norwell, MA: Kluwer, 1990.

Raz, Joseph. "Government by Consent." In J. Roland Pennock and John W. Chapman, eds., *Nomos XXIX, Authority Revisited.* New York University Press, 1987.

———. "On the Nature of Rights." In Morton E. Winston, ed., *The Philosophy of Human Rights.* Belmont, CA: Wadsworth, 1989.

———, ed. *Authority.* New York: NYU Press, 1990.

Reinhardt, Mark. "Look Who's Talking: Political Subjects, Political Objects, and Political Discourse in Contemporary Theory." *Political Theory* 23:689–719, 1995.

Rescher, Nicholas. *Risk.* Washington DC: University Press of America, 1983.

Ricci, David. *The Tragedy of Political Science.* New Haven, CT: Yale University Press, 1984.

Richards, David A. J. "Human Rights, Public Health, and the Idea of Moral Plague." *Social Research* 55:491–528, 1988.

Ripstein, Arthur. "Equality, Luck and Responsibility." *Philosophy and Public Affairs* 23:3–23, 1994.

Robinson, Michael. *Over the Wire and on TV.* New York: Russell Sage, 1983.

Rosenblum, Nancy, ed. *Liberalism and the Moral Life.* Cambridge, MA: Harvard University Press, 1989.

Rourke, John T., Richard P. Hiskes, and Cyrus Ernesto Zirakzadeh. *Direct Democracy and International Politics.* Boulder, CO: Lynne Rienner, 1991.

Ruben, David-Hillel. "Tacit Promising." *Ethics* 83:71–9, 1982.

Rudner, Richard S. *Philosophy of Social Science.* Englewood Cliffs, NJ: Prentice-Hall, 1966.

Sandel, Michael J. *Liberalism and the Limits of Justice.* Cambridge: Cambridge University Press, 1982.

Sapolsky, Harvey M. *Consuming Fears.* New York: Basic, 1986.

Schaar, John. *Legitimacy in the Modern State.* New Brunswick, NJ: Transaction Press, 1981.

Schiller, Herbert. *Communication and Cultural Domination.* White Plains, NY: International Arts and Science Press, 1976.

Schwing, R. C. and W. A. Albers, eds. *Societal Risk Assessment: How Safe is Safe Enough?* New York: Plenum, 1980.

Scott, John T., and Vickie B. Sullivan. "Patricide and the Plot of *The Prince*: Cesare Borgia and Machiavelli's Italy." *American Political Science Review* 88:887–900, 1994.

Settle, D. M., and C. C. Patterson. "Lead in Albacore: Guide to Lead Pollution in Americans." *Science* 207:1167–76, 1980.

Sheffler, Israel. *Science and Subjectivity*. Indianapolis: Hackett, 1982.

Shklar, Judith. *Men and Citizens*. London: Cambridge University Press, 1969.

Shrader-Frechette, K. S. "Consent and Nuclear Waste Disposal. *Public Affairs Quarterly* 7:363–77, 1993.

———. *Risk Analysis and Scientific Method*. Dordrecht: D. Reidel, 1985.

———. *Risk and Rationality*. Berkeley: University of California Press, 1991.

Simon, Michael Arthur. "Causation, Liability and Toxic Risk Exposure." *Journal of Applied Philosophy* 9:35–44, 1992.

Singer, Peter. "Famine, Affluence, and Morality." *Philosophy and Public Affairs* 1:229–43, 1982.

Slovic, Paul. "Perception of Risk: Reflections on the Psychometric Paradigm." In Sheldon Krimsky and Dominic Golding, eds., *Social Theories of Risk*. Westport, CT: Praeger, 1992.

———. "Public Perceptions of Risk." *Risk Management* March, 1992.

Slovic, Paul, James H. Flynn, and Mark Layman. "Perceived Risk, Turst, and the Politics of Nuclear Waste." *Science* 254:1603–7, 1991.

Smiley, Marion. *Moral Responsibility and the Boundaries of Community*. Chicago: University of Chicago Press, 1992.

Starr, Chauncey. "Social Benefit versus Technological Risk. *Science* 165:1232–38, 1969.

Stern, Paul. "The Rule of Wisdom asnd the Rule of Law in Plato's *Statesman*." *American Political Science Review* 91:264–76, 1997.

Strawson, P. F. *Freedom and Resentment and Other Essays*. London: Methuen, 1974.

Sumner, L. W. *Abortion and Moral Theory*. Princeton, NJ: Princeton University Press, 1981.

———. *The Moral Foundation of Rights*. Oxford: Clarendon Press, 1987.

Sunnstein, Cass. "Constitutional Politics and the Conservative Court." *American Prospect* 1:51–61, 1990.

Taylor, Charles. *Sources of the Self: The Making of Modern Identity*. Cambridge, MA: Harvard University Press, 1989.

Thompson, Michael. "Aesthetics of Risk: Culture or Context?" In R. C. Schwing and W. A. Albers, eds. *Societal Risk Assessment: How Safe is Safe Enough?* New York: Plenum, 1980.

———. "To Hell With the Turkeys! A Diatribe Directed at the Pernicious Trepidity of the Current Intellectual Debate on Risk." In Douglas Maclean, ed., *Values at Risk*. Lanham, MD: Rowman and Littlefield, 1986.

Thompson, Dennis F. "Moral Responsibility of Public Officials: The Problem of Many Hands." *American Political Science Review* 74:905–16, 1980.

Thomson, Judith Jarvis. *The Realm of Rights*. Cambridge, MA: Harvard University Press, 1990.

———. *Rights, Restitution, and Risk*. Cambridge, MA: Harvard University Press, 1986.

Tuana, Nancy, ed. *Feminism and Science*. Bloomington: Indiana University Press, 1989.

Unger, Roberto M. *The Critical Legal Studies Movement*. Cambridge, MA: Harvard University Press, 1986.

Verba, Sidney, Kay Lehman Schlozman, and Henry E. Brady. *Voice and Equality: Voluntarism in American Politics*. Cambridge, MA: Harvard University Press, 1995.

Waldron, Jeremy. *Nonsense Upon Stilts: Bentham, Burke and Marx on the Rights of Man*. London: Methuen, 1987.

Wall, James A., Jr. "Mediation: An Analysis, Review, and Proposed Research." *Journal of Conflict Resolution* 25:157–80, 1981.

Walzer, Michael. "The Communitarian Critique of Liberalism." *Political Theory* 18:6–23, 1990.

———. *Obligations: Essays on Disobedience, War, and Citizenship*. New York: Clarion, 1970.

Warren, Mark E. "Deliberative Authority and Democracy." *American Political Science Review* 90:46–60, 1996.

Williams, Bernard. *Moral Luck*. Cambridge: Cambridge University Press, 1981.

Winner, Langdon. *Autonomous Technology*. Cambridge, MA: MIT Press, 1977.

Winston, Morton E., ed. *The Philosophy of Human Rights*. Belmont, CA: Wadsworth, 1989.

Wolff, Robert Paul. *In Defense of Ararchy*. New York: Harper, 1970.

———. *The Poverty of Liberalism*. Boston: Beacon, 1968.

Wolin, Sheldon. *Politics and Vision*. New York: Little, Brown, 1960.

Wynne, Brian. "Risk and Social Learning: Reification to Engagement." In Sheldon Krimsky and Dominic Golding, ed., *Social Theories of Risk*. Westport, CT: Praeger, 1992.

———. "Technology, Risk, and Participation: On the Social Treatment of Uncertainty." In J. Conrad, ed. *Science, Technology, and Risk*. New York: Academic Press, 1980.

———., ed. *Risk Assessment and Hazardous Waste Management*. London: Springer-Verlag, 1987.

Young, Iris. *Justice and the Politics of Difference*. Princeton, NJ: Princeton University Press, 1990.

Zimmerman, Andrew D. "Toward a More Democratic Ethic of Technological Governance." *Science, Technology, & Human Values* 20:86–107, 1995.

Index